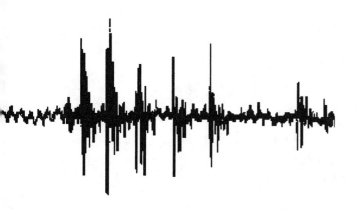

RETURN OF
GONZO GIZMOS

MORE PROJECTS & DEVICES
TO CHANNEL YOUR INNER GEEK

SIMON QUELLEN FIELD

CHICAGO
REVIEW
PRESS

Library of Congress Cataloging-in-Publication Data
Field, Simon (Simon Quellen)
Return of gonzo gizmos : more projects & devices to
channel your inner geek / Simon Quellen Field.
    p. cm.
Includes index.
ISBN-13: 978-55652-610-7
 ISBN-10: 1-55652-610-5
1. Electronic apparatus and appliances—Design
and construction—Amateurs' manuals. 2. Science—
Experiments. I. Title.
TK9965.F475 2006
621.381—dc22   2005034276

Cover and interior design: Laura Lindgren

Published by Chicago Review Press, Incorporated
814 North Franklin Street
Chicago, Illinois 60610
ISBN-13: 978-1-55652-610-7
ISBN 1-55652-610-5
Printed in the United States of America
5   4   3   2   1

Contents

➊ CHEMISTRY

➋ AERODYNAMICS

➌ ELECTRICITY AND MAGNETISM

4 COMPUTERS AND ELECTRONICS

5 MATHEMATICS

6 BIOLOGY

Introduction

I love unusual gizmos, especially those I can make myself.

Most people who enjoy gizmos freely admit that they are toys. The devices might have some beneficial use, but a large part of why people love them is that they are *fun*, plain and simple.

Sometimes just being small makes something cool. A tiny rocket engine—like the one in this book—is fun partly just because it is so tiny. A microscope is interesting because it lets you see things that have always been invisible to your naked eye.

Another reason a small rocket engine is great is because of what I call "action at a distance": being in one place and having an effect in another, distant place. This is why slingshots and bows and arrows are popular, as well as snowballs and walkie-talkies. In this book you will find very simple radio transmitters and laser transmitters that can send your messages over long distances.

Building your own gizmos is part of the fun as well. This is especially true when the device is mysterious or challenging, such as anything to do with lasers. Building your own radio transmitter or receiver (or both) *sounds* impressive, but often what is most impressive is just how easy it can be to build something that most people would think impossible. You can build the radio transmitters and receivers in this book in just a few minutes, even if you have never done anything with electronics before. The laser transmitter and receiver are simple as well.

Magnets are especially fun because few other things possess that invisible force that can be felt in your hands but not perceived by any other senses. Using magnets to shoot or propel objects adds that great "action at a distance" component.

Sometimes the name of a toy adds to the fun. Telling people you built your own rail gun, hydrogen fuel cell, or geodesic dome (all projects in this book) sounds really high-tech (and perhaps a little frightening).

And some gizmos are fun because they do amazing things. You can listen in on the electrical navigation signals of a weird little fish using a gizmo in this book, and open your ears to a part of the world you probably never knew existed. You can float a wooden ball in midair, balanced on an invisible breath by an unseen force. You can paint a safe, nontoxic liquid metal alloy onto a piece of glass to create a homemade mirror.

With almost every one of these projects, the first words after "Wow!" are usually "How does it do that?" All of these gizmos come with an explanation of the scientific principles that make them work. Using simple language and illustrations, each gizmo is described so that even the youngest audience can understand, but with enough detail to satisfy the engineer, teacher, or inventor whose mind is now taking the project in entirely new directions.

All, of course, in the name of fun.

Thoughts on Safety

This book is not for people who don't like to think.

It is not for people who like to smash their fingers with a hammer, or burn their house down.

Don't give this book to people like that.

There aren't a lot of really dangerous projects in this book, but putting a little thought into basic safety is always a good idea. Several of the projects in the book deal with soldering irons and other hot things. Don't put them into your mouth or your ear. Use lead-free solder if you plan to eat your radio. Don't aim the rail gun at the parakeet. Don't play with matches in the house, in the garage, or in a field of dry grass. If you like to run with scissors, this is not the book for you.

Science projects are fun to show off because they make you look smart. Don't ruin the effect by bleeding all over the equipment or setting your hair on fire. Wearing safety goggles makes any project look more impressive, and protects your eyes to boot.

Take good care of the electric fish. They like to jump out of the tank, so keep it covered. Electric fish don't work once they have dried out. Give your fish a name and talk to it during the project, so you will be motivated to keep your little friend safe. Houdini is a good name for a critter that performs escape stunts at great peril to its life.

Keep a fire extinguisher on hand at all times. Like safety glasses, it adds a sense of danger and excitement to the project and can keep a simple mistake from becoming a tragedy.

Most of all, use your head, and slow down a little if you feel you are getting overly excited. This is supposed to be fun.

CHEMISTRY

Hydrogen Fuel Cell

A fuel cell is a device that converts a fuel such as hydrogen, alcohol, gasoline, or methane directly into electricity. A hydrogen fuel cell produces electricity without any pollution; pure water is the only by-product.

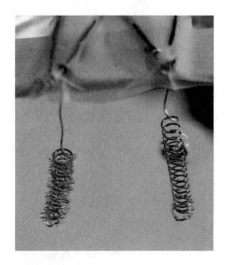

Hydrogen fuel cells are used in spacecraft and other high-tech applications that need a clean, efficient power source.

You can make a hydrogen fuel cell in your kitchen in about 10 minutes, and demonstrate how hydrogen and oxygen can combine to produce clean electrical power.

SHOPPING LIST

- 1 foot of platinum-coated nickel wire or pure platinum wire (available from science supply stores, or at www.scitoys.com)
- Popsicle stick or similar small piece of wood or plastic
- 9-volt battery clip
- 9-volt battery
- Transparent sticky tape
- Glass of water

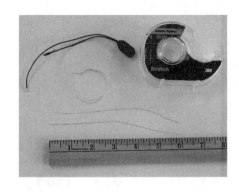

TOOLS

- Volt meter

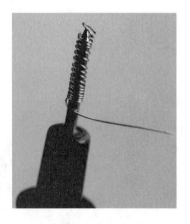

First, cut the platinum-coated wire into two 6-inch-long pieces, and wind each piece into a little coiled spring that will serve as the electrodes in the fuel cell. The wires in the photo were wound on the end of the test lead of a volt meter, but a nail, an ice pick, or a coat hanger works nicely as a coil form.

Next, cut the leads of the battery clip in half and strip the insulation from the cut ends. Then twist the bare wires onto the ends of the platinum-coated electrodes, as shown in the photo at right. The battery clip and two wires will also be attached to the electrodes. These will later be used to connect to the volt meter.

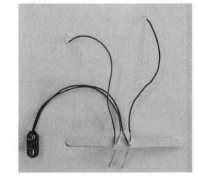

Tape the electrodes securely to the Popsicle stick. Then tape the Popsicle stick to the glass of water, so that the electrodes dangle in the water for nearly their entire lengths. The twisted wire connections must stay out of the water, so only the platinum-coated electrodes are in the water.

Now connect the red wire to the positive terminal of the volt meter, and the black wire to the negative (or common) terminal of the volt meter. The volt meter should read 0 volts at this point, although a tiny amount of voltage may show up, such as 0.01 volt.

Your fuel cell is now complete.

To operate the fuel cell, touch the 9-volt battery to the battery clip (don't actually clip it on; you will only need it for a second or two).

Touching the battery to the clip causes the water at the elec-

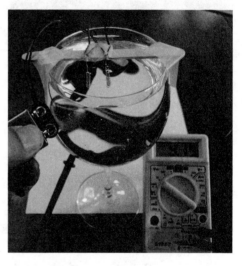

trodes to split into hydrogen and oxygen, a process called electrolysis. You can see the bubbles form at the electrodes while the battery is attached. Bubbles of hydrogen will cling to one electrode, and bubbles of oxygen will cling to the other.

Now remove the battery. If you did not use platinum-coated wire, you would expect the volt meter to read 0 volts again, since there would be no battery connected. But the platinum acts as a catalyst, allowing the hydrogen and oxygen to recombine—the hydrolysis reaction reverses! Instead of putting electricity into the cell to split the water, the hydrogen and oxygen combine to

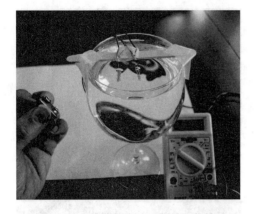

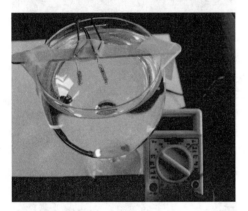

make water again, *and produce electricity.*

You should initially get a little more than 2 volts from the fuel cell. As the bubbles pop, dissolve in the water, or get used up by the reaction, the voltage drops, quickly at first, then more slowly.

After a minute or so, the voltage declines much more slowly, as most of the decline is now due only to the gases being used up in the reaction that produces the electricity.

In this project you have stored the energy from the 9-volt battery as hydrogen and oxygen bubbles. You could, instead, bubble hydrogen and oxygen from some other source over the electrodes and generate electricity. Or you could produce hydrogen and oxygen during the day from solar power, store the gases, and then use them in the fuel cell at night. You could also store the gases in high-pressure tanks in an electric car, and generate the electricity the car needs from a fuel cell.

☻ WHY DOES IT DO THAT? ☻

There are two things going on in this project—the electrolysis of water into hydrogen and oxygen gases, and the recombining of the gases to produce electricity. Let's look into each step separately.

The electrode connected to the negative side of the battery has electrons that are being pushed by the battery. Four of the electrons in that electrode combine with four water molecules (H_2O). The four water molecules each give up a hydrogen atom, to form two molecules of hydrogen (H_2), leaving four negatively charged ions of OH^-.

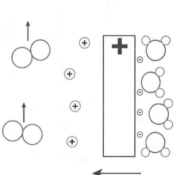

The hydrogen gas bubbles up from the electrode, and the negatively charged OH^- ions migrate away from the negatively charged electrode.

At the other electrode, the positive side of the battery pulls electrons from the water molecules. The water molecules split into positively charged hydrogen atoms (single protons), and oxygen molecules. The oxygen molecules bubble up, and the protons migrate away from the positively charged electrode.

The protons eventually combine with the OH^- ions from the negative electrode, and form water molecules again.

The Fuel Cell

When you remove the battery, the hydrogen molecules clinging as bubbles to the electrode break up due to the catalytic action of the platinum, forming positively charged hydrogen ions (H^+, or protons) and electrons.

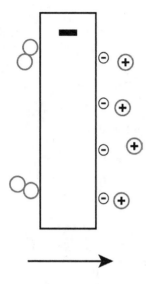

At the other electrode, the oxygen molecules stuck in bubbles on the platinum surface draw electrons from the metal, and then combine with the hydrogen ions in the water (from the reaction at the other electrode) to form water.

The oxygen electrode has lost two electrons to each oxygen molecule. The hydrogen electrode has gained two electrons from each hydrogen molecule. The electrons at the hydrogen electrode are attracted to the positively charged oxygen electrode. Electrons travel more easily in metal than in water, so the current flows in the wire instead of the water. In the wire, the current can do work, such as illuminating a light bulb or moving a volt meter.

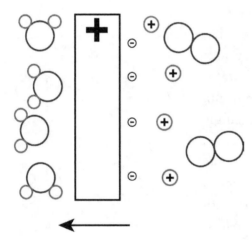

Mirrors and Liquid Metal Alloys

Suppose you had a metal alloy that had the advantages of liquid mercury, but without the toxic effects?

You could make your own barometers and thermometers without having to worry about calling in a hazardous-materials team to clean up any accidents—you could simply wipe up the mess with a paper towel. You wouldn't have to worry about breathing toxic mercury fumes, but you could still make neat little electric motors that dip into liquid metal to make their electrical connections.

Suppose further that the metal would stick to glass, so you could paint it on glass to make your own mirrors. Or that it would stick to paper so you could draw your own electric circuits in it.

SHOPPING LIST

- ꜫ Vial of gallium or gallium alloy (available from www.scitoys.com)
- ꜫ Cotton swabs
- ꜫ Glass microscope slide
- ꜫ Glass slide slipcover (optional)

TOOLS

- ꜫ Laser pointer (optional)

In the photo to the left are two small vials of liquid metal. The vial on the right contains gallium, an element that melts at 85.57°F (29.76°C). The vial on the left is an alloy that contains gallium, indium, and tin, and melts at −4°F (−20°C).

The gallium is liquid because the bottle was stored in a shirt pocket, next to a warm body. At room temperature it is a solid.

Because gallium expands when it solidifies (unlike most metals), the vials are only filled halfway. To get the solid metal out of the vial, simply warm it up in a cup of hot water until it melts.

Fun Projects with Liquid Metal

One fun thing you can do right away with the liquid metal alloy is make your own mirror. All it takes is a piece of glass and a cotton swab.

Dip the cotton swab in the vial, and twirl it around to coat it with the liquid metal alloy.

Now rub the coated swab on a glass microscope slide. The metal sticks to the glass, and makes an opaque reflective coating.

In the third photo on the left, I am holding the new mirror so that it reflects the view of the trees outside my window. The camera is focused on the window, so the trees and my hand are out of focus.

Being able to make your own mirrors is helpful when you need one that might be hard to find in stores. For example, I once needed a small lightweight mirror to glue to a speaker, so I could bounce a laser beam off of the speaker and have the music wiggle the mirror, which in turn made a pattern on the wall.

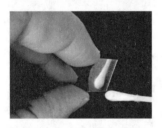

I used the liquid metal to coat a thin glass cover slip for a microscope slide.

The resulting mirror was very lightweight, and yet stiff, so it would remain flat while being bounced around by the speaker.

When it was glued onto the speaker and the

music was turned on, the laser created a light show on the wall. Using two speakers, and bouncing the light off one and then off the other, I made a computer sound file that used both stereo channels to draw pictures on the wall.

Uses for Liquid Metal

There are lots of things you can do with liquid metal:
* Make thermometers
* Make barometers
* Make tilt meter seismographs
* Make nonconductive objects conductive
* Make electrodes that conform to varying surfaces
* Experiment with magnetohydrodynamics
* Conduct high-energy sound
* Replace mercury in spinning telescope mirrors

To keep the surface shiny, coat it with a diluted solution of hydrochloric acid or a thin layer of mineral oil. Both will prevent the slow oxidation of the metal that occurs over time.

WHY DOES IT DO THAT?

Gallium is an element, atomic number 31, right below aluminum and just to the right of zinc in the periodic table of the elements. It starts out with a very low melting point, but when chemists add other elements it can achieve an even lower melting point.

Just below gallium in the periodic table is indium (element 49). Just to the right of indium is tin (element 50). When these elements are combined, their atoms bind together into a compound. The molecules of that compound do not bind to one another as strongly as the atoms of the original metals bound to each other. This lowers the melting point.

There are many ways to combine the three metals:

Compound	Percentages	Grams Ga	Grams In	Grams Sn
$Ga_{14}In_3Sn_2$	62.65% Ga, 22.11% In, 15.24% Sn	97.6122	34.4454	23.742
$Ga_{17}In_4Sn_2$	62.98% Ga, 24.40% In, 12.62% Sn	118.529	45.9272	23.742
$Ga_{22}In_5Sn_3$	62.25% Ga, 23.30% In, 14.45% Sn	153.391	57.409	35.613
$Ga_{25}In_5Sn_4$	62.43% Ga, 20.56% In, 17.01% Sn	174.308	57.409	47.484
$Ga_{25}In_6Sn_3$	62.52% Ga, 24.71% In, 12.77% Sn	174.308	68.8908	35.613

. . . and so on.

Each combination will have a slightly different melting point. Which do you think has the lowest melting point? This might make a good science fair experiment.

A mixture of 76 percent gallium and 24 percent indium melts at 61°F (16°C). Both gallium and this combination can be super-cooled. That means that once melted, they can stay liquid even though they are cooled well below their melting points. Eventually a small crystal forms, and starts the whole batch solidifying, but small amounts can be kept supercooled for quite a while.

The gallium-indium alloy is more reflective than mercury, and is less dense, so it is being explored as a replacement for mercury in spinning liquid mirrors for astronomical telescopes.

When gallium is exposed to air, a thin layer of gallium oxide forms on the surface, just like what happens with aluminum, the metal just above it in the periodic table. This allows gallium alloys to wet almost any material, which means that instead of beading up, they spread out over the surface. This property makes gallium alloys good for making mirrors, and for coating objects to make them conductive.

In the same way that mercury alloys with other metals to make amalgams, gallium also alloys with other metals. When a small drop of gallium is placed on

aluminum foil, for example, it will combine with the aluminum to make a liquid
with a crusty surface, as in the photo on
page 10.

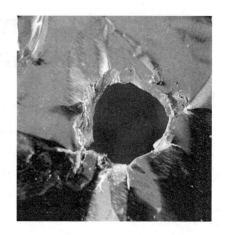

The alloy eventually combines with
all of the aluminum, dissolving a hole
in it.

If a drop of water is added to the
resulting bead of liquid metal, the water
combines vigorously with the aluminum,
making a hot solution of caustic alumi-
num hydroxide. What is left is the original drop of gallium with a
tiny amount of aluminum dissolved in it. (Don't put that drop
back in the bottle; it will contaminate the rest of the gallium.)

This experiment can be done with either the gallium or the
gallium-indium-tin alloy.

Homemade Ice Cream
Without an Ice Cream Maker

Here is a way to make fresh homemade ice cream by hand in much less time than it normally takes with a home ice cream maker.

The ice cream freezes much faster because you prepare each serving in its own batch of ice and salt. Also, each serving is prepared in a thin, flat container, so the ice and salt can contact more of the ice cream at once.

SHOPPING LIST

- ½ cup sugar
- ¼ teaspoon salt
- 1 cup milk
- 3 beaten egg yolks
- 1 tablespoon vanilla extract
- 2 cups chilled whipping cream
- 2 cups fresh strawberries and extra ½ cup sugar (optional)
- Zip-lock sandwich bags
- Gallon-size zip-lock sandwich bags
- Ice cubes
- Rock salt

TOOLS

- Measuring cup
- Measuring spoons
- Mixing bowl
- Double boiler
- Wooden spoon

The cooking phase can be done a day or two ahead of time, so no one has to wait for the really fun part.

Set out all of the ingredients so everything is within easy reach.

Put the sugar, salt, and milk into the top pan of a double boiler. The water in the bottom of the double boiler will boil, and the temperature will never rise above the boiling point of water. This ensures that even if you get distracted you won't overcook the mixture.

Stir the 3 beaten egg yolks into the milk and sugar.

Cook the mixture over boiling water until you see bubbles forming around the edges.

The mixture is done when it is thick enough to coat the spoon.

Let the mixture cool to room temperature. When it has cooled, stir in the vanilla extract and the heavy whipping cream.

The fun part is about to begin. This is where having a bunch of kids around to help is really nice.

Pour a cup of the ice cream mixture into a plastic zip-lock sandwich bag.

Zip the bag, put it inside another sandwich bag for safety, and zip that one closed as well.

Fill a gallon-size food storage zip-lock bag about one third full of ice cubes. Add 1 cup of salt (we used rock salt, but any kind will do).

Zip the large bag closed, and wrap it in a towel to keep fingers from getting too cold.

Make a bag for everyone (this recipe will make enough for three or four servings, and you can double or quadruple the recipe if you're going to have a party).

Now have each person squish the little bag around in the salt and ice, making sure that the ice contacts the bag as much as possible, and that the bag gets lots of kneading, to keep the ice crystals tiny, so the ice cream will come out very smooth.

The kneading stage takes 10 minutes. You can let the ice cream sit in the ice for another 5 minutes if you like firmer ice cream, although continuing to knead it for the extra 5 minutes is also perfectly fine if you're having fun.

You will know the ice cream is done by feeling the mixture become a paste instead of a liquid. When you take the little bag out of the ice, wipe off the salt water, and then remove the outer bag carefully so you don't get salt in the ice cream. The bag will stand up in the bowl because it has turned into a frozen paste.

You can spoon the ice cream into a bowl if you like, or just eat it out of the bag.

If you like strawberry ice cream, mash 2 cups of strawberries with ½ cup sugar, and add ½ cup to each small bag before closing it up and putting it in the ice.

The result is an amazingly delicious homemade ice cream.

WHY DOES IT DO THAT?

For ice to melt, it has to get heat from something. In this ice cream project, it gets the heat from the ice cream mixture (and from your hands, which is why they get cold while holding the bag). When the ice is melting, it is at 32°F (0°C).

When ice melts, the surface of the ice is wet. At the surface, there is solid ice on one side, and liquid water on the other. The boundary surface is exactly at the freezing point. This means that some water molecules are leaving the ice and moving into the water, but it also means that some liquid water is refreezing onto the ice. The system is in equilibrium when the rate of melting is equal to the rate of freezing, and this happens at 0°C.

At equilibrium, the heat lost by the water as it freezes is equal to the heat gained by the ice as it melts.

Because plain ice can only barely cool something to the freezing point of water, something must be done to make it much colder than that, since the ice cream mixture freezes at a lower temperature than water.

Salty water freezes at a lower temperature than plain water. But your ice is made of plain water, so it melts at 0°C. Since the ice keeps melting, but the water no longer freezes (because there is only salt water, which doesn't freeze at 0°C), the temperature goes down.

The heat gained by the ice as it melts is no longer offset by the heat given up by freezing water (since the water is no longer freezing back onto the ice). The heat gain has to come from somewhere else. It comes from the ice cream and your hands.

The sodium and chlorine in the salt split apart into charged ions, and these ions attract water molecules to form weak chemical bonds.

The resulting compound has a freezing point of −5.98°F (−21.1°C). This is 21.1°C (37.98°F) colder than ice.

If you put salt on the icy pavement of a sidewalk or a road, the ice mixes with the salt and the mixture of the two solids (ice and salt) produces a liquid, but the sidewalk actually gets colder than it was before.

If you add a different chemical to the ice, such as calcium chloride, you get an even lower temperature: −20°F (−29°C).

2

AERODYNAMICS

Bernoulli Ball

You may have seen beach balls balancing on vacuum cleaner exhaust hoses. You can make a smaller version by balancing a ping-pong ball on a stream of air from a blow dryer.

The toy you will make in this section uses the same principles as the beach ball and the ping-pong ball. A small ball of balsa wood floats on a stream of air you blow through a bent tube. The ball has a small wire hook stuck through it. The object of the game is to hook the ball on a hoop attached to the tube.

SHOPPING LIST

- ↄ 6 inches of clear plastic tubing, ½ inch in diameter
- ↄ 1 inch of clear plastic tubing, ⅛ inch in diameter
- ↄ Plastic glue
- ↄ Balsa wood ball, ½ inch in diameter

TOOLS

- ☐ Drill
- ☐ Needle-nosed pliers

Start with a clear plastic tube, ½ inch in diameter. These can be found in hobby stores or hardware stores. Drill a single ⅛-inch hole ½ inch from the end of this large plastic tube.

Glue a smaller clear plastic tube over the hole drilled in the larger tube. The idea is to make a tube that bends at a right angle, so the air coming from your mouth comes out vertically about six inches from your face.

Close the far end of the long tube by gluing on a circle of plastic cut to fit.

Push a stiff wire hook through the balsa wood ball, so that about an inch and a half sticks out the bottom, and an inch sticks out the top. Bend the top into a hook about ½ inch above the ball. When you are finished, the ball should not slide on the wire.

To make the hook, use something that doesn't rust, such as stainless steel or brass. You can see the rust stains on the plas-

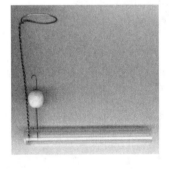

tic tube in the version in the photos, which was caused by warm, moist air.

Form the hoop by bending the wire in half around a broomstick and then twisting the ends to create the handle of the loop. Insert this handle into a hole drilled near the far end of the tube.

The completed toy looks like the illustration at left.

Playing Around

To operate the toy, place the straight wire of the ball in the upright tube as shown on the previous page. Take a very deep breath (you'll need all the air you can get). Blow through the open end of the large tube hard enough to raise the ball up to the level of the hoop.

The ball will now dance around on top of the stream of air and act very erratically. The trick is to control the height of the ball so when the hook rises above the hoop, you can let the ball down gently and hook the hoop. This is not easy, and it usually takes several breaths. It is fun to watch the expressions on their faces as people try to hook the loop on their last bit of air.

WHY DOES IT DO THAT?

To explain what is happening with the balancing ball, it is helpful to look at other, simpler experiments.

Some Simple Experiments

When you hold your hand out the window of a moving car, you can make the wind lift your hand by tilting your hand at an angle to the wind. Notice that the force you feel that lifts your hand seems to press on the bottom of your hand. If there is any force acting on the top of your hand trying to suck it upward, it is definitely smaller than the force on the bottom of your hand, which is pushing it upward. (Try sucking on the back of your hand using your mouth. It is easy to detect even small amounts of suction. The amount needed to lift your hand would definitely be noticed.)

It is easy to picture the molecules of air hitting the bottom of your hand and pushing it up. It is also easy to determine that the air that is hitting your hand is bouncing off your hand and going downward. You can hold your hand in such a way as to make the air bounce off your hand and hit your face. When you do this, your hand is pushed in a direction away from your face.

Whichever way your hand directs the air, the air pushes your hand in the opposite direction. The amount of force your hand feels depends on the amount of air that is pushed in the opposite direction. If you hold something larger than your hand out the window of the car, you will feel a larger force, since more air is being moved. The amount of force your hand feels also depends on the speed of the car. The faster the wind, the larger the force. And, too, the density of the air affects the force you feel. If you put your hand in the water when you travel by boat, you feel a much larger force, even if the boat is going much slower than the car.

Lift and Drag

Notice that you feel the largest lifting force on your hand when it is held at a 45-degree angle. This angle causes the wind to bounce off your hand straight down. But you also feel a force pushing your hand backward, away from the direction of travel. This is because you are stopping the air molecules from traveling backward, and are making them go down instead. The backward force you feel is equal to the upward force.

The backward force (the force resisting forward motion) is called *drag*, and it is easy to see that you cannot get lift without drag. You cannot change the direction of the wind without feeling its resistance to change.

There is another type of drag that is important to understanding how the ball balances on the stream of air, or how airplanes fly, or how sailboats sail into the wind. Look again at your hand moving through water. If your hand moves very slowly, it will not stir up the water very much behind it. If you move your hand more quickly, you will see little whirlpools form behind it. It takes force to cause all this water motion, and that force is felt as added resistance to the movement of your hand. If you could somehow streamline your hand to get the most movement of water in the direction you want, while causing the least stirring up of the water, you would have less drag for a given lift, and you would have a more efficient wing, propeller, or sail.

Viscosity and Drag

In water, you have to move very slowly to avoid causing whirlpools (also known as *vortices*). In air, you can move a little faster without stirring up the air in the same way (you can see this if you use smoke to make the vortices visible). In a jar of honey, you have to move very slowly. The difference is *viscosity*.

Viscosity is a property of fluids (like air or water). It is the ability of the fluid to resist changes in its shape that do not change its volume.

Viscosity is caused by interactions between the molecules of the fluid. It is the transfer of *momentum* from one part of the fluid to another part.

In a gas such as air, viscosity is caused almost entirely by collisions between molecules. The faster the molecules are moving, the more effective is the transfer of momentum. In a hot gas, the molecules are moving faster than in a cold gas, so the viscosity is higher. A small part of viscosity in air is caused by attractive forces between molecules. These forces, which are known as *Van Der Waals forces*, are much larger in water and honey, and they play a bigger part there. In air, Van Der Waals forces are small enough to ignore when explaining lift and drag.

It is useful to compare the momentum forces in a fluid to the viscous forces. The ratio of the two is called the *Reynolds number*. It is defined as the density times the velocity times the length (width of your hand) all divided by the viscosity.

If the viscosity is low, as it is in air, or the speed is slow, then your hand does not transfer momentum to very much air. Almost all of the momentum that is transferred goes into moving the air downward, and very little goes into stirring it up. As the speed increases, or the viscosity increases, more of the momentum is transferred to the air above, below, and behind your hand, where it is wasted as extra drag.

Curved Surfaces and the Coanda Effect

If you limit the angle that the fluid has to turn as it passes over a wing, then you can limit the rotation of the fluid. The fluid won't spin around as much if you don't kick it very much. If you put a cylinder in the water, you will see a lot of vortices created as it moves. If you shape it like a fish or a teardrop, where the trailing edge gradually tapers to a point, then the water does not have to turn as sharply, and so it does not spin as much and smaller vortices are produced.

Now you can see why some wings are curved on the top. By gradually letting the air fill the empty space behind the wing, you limit the amount of spin you impart to the vortices. This limits the drag on the wing.

The tendency of a fluid to follow a curved surface is called the Coanda effect.

Notice what is happening in the following photograph of smoke pulses flowing over an airplane wing. The air slows down as it gets close to the wing. The Bernoulli principle says that this slower moving air will appear to the wing to have a higher pressure than faster moving air. What keeps this high pressure from pushing the wing down is the fact that it happens on the bottom of the wing as well, so it is balanced.

Note also that the air does not speed up as it moves over the curved top of the wing, but it does slow down as it encounters the tilted bottom of the wing. You can measure the different pressures on the top of the wing and on the bottom, and the difference is lift. You get the same value for lift whether you look at the mass of air moving downward or the pressure difference

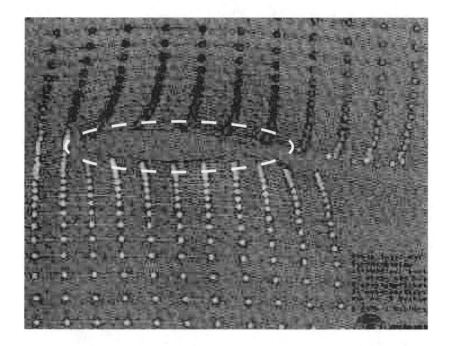

between the top of the wing and the bottom, because they are two different ways of looking at the same thing.

Notice that the viscosity of the fluid causes it to follow the shape of the wing. In a gas, the viscosity is the result of collisions between molecules. The molecules above the wing are constantly bumping into one another. As the wing sweeps away the molecules in front of it and pushes them downward, it leaves an empty space behind it. The air above this empty space expands into it due to the collisions of the molecules.

Picture the wing as having two springs attached to it, one pushing down on the top of the wing, and one pushing up on the bottom of the wing. If you move the wing down, you compress the bottom spring, and the top spring expands because you are no longer pushing on it as hard as you were before. The springs are the air molecules bouncing against the wing and each other. You are moving two masses of air in the downward direction. The air above the wing moves down as well as the air under the wing.

Before the wing moved the air, the air on the bottom was holding up the air on the top. In order to do this, there must have been a force pushing upward. The wing moving through the air must not only accelerate the molecules in a downward direction, but it must overcome the upward force that is holding up the air above the wing. You can think of this upward force as helping to hold up the wing now instead of holding up the air above the wing. It thus adds to the lift on the wing. The air above the wing now falls into the empty space behind the wing. It also falls past the wing as the wing moves out of the way, and you can measure the amount of air that is moving down and see that it matches the lift on the wing as expected.

Back to Why the Ball Balances . . .

You are now finally ready to see why the ball balances on the stream of air.

To balance, the ball must see a force that tends to center it on the air stream when it strays. From our discussion above, you would expect to see two things happen if this force exists. You would see air moving in the opposite direction of the ball's motion. You would also expect to see a higher pressure on the side of the ball opposite the air stream, and a lower pressure on the side facing the air stream.

Picture the air stream grazing the ball on the left side. The curve of the ball is fairly gentle, and causes the air to follow the curve. As the air follows the curve, it moves away from the stream of air. If the air is moving away from the stream of air, whatever caused it to move (in this case, the ball) must feel a force toward the stream of air.

This air on the left side is moving faster than the air on the right side (which isn't moving). As the air moves past the ball, it sweeps aside air molecules that were moving toward the ball, and would have hit the ball if they had not been moved aside. The pressure on that side of the ball is thus lower.

On the right side of the ball the air is not moving, so the pressure has not changed. The pressure on the left is lower than the pressure on the right, so the ball moves toward the stream of air.

No matter which direction the ball is deflected, it is attracted to the center of the air stream and stays balanced.

The Bernoulli principle states that the pressure on the ball from the side toward the moving air is less than that on the side where the air is still. This is why I call the toy the Bernoulli Ball. Notice that what actually moves the ball is the recoil of lots more tiny air molecules on the right side of the ball than on the left. I could call the toy the Newton Ball, but that lacks alliteration. Looking at it another way, you see air moving away from the stream of air as it follows the curve of the ball. So I could also call the toy the Coanda Ball, although I prefer to think of the Coanda effect as the *result* of lift, not the *cause*.

Vacuum Pump

Some interesting things can be accomplished once you are free of the pesky atmosphere we live in. Feathers drop like rocks. Bells go silent. Water boils at room temperature. Balloons inflate themselves. Frozen foods dry out but keep their shape.

In many cases, all you need is a little bit of vacuum to help dry something out, or move a liquid through a tube, or remove gases from a liquid.

In this project, you will make a vacuum pump from common inexpensive plumbing parts. You will then use it to inflate marshmallows to twice their size, remove the air from them, and then shrink them down to a wrinkled, rubbery candy with the density of a gumdrop.

If all of the materials are at hand, the pump can be put together in about a half hour.

SHOPPING LIST

- 4 feet of 1-inch-diameter Schedule 40 PVC pipe
- 4 feet of ½-inch-diameter Schedule 40 PVC pipe
- 1 foot of ¾-inch-diameter Schedule 80 (thin wall) PVC pipe
- 1-inch-diameter PVC check valve
- ½-inch-diameter PVC T-fitting
- 1 foot (or so) of clear vinyl hose
- ½-inch barb fitting to match hose
- 1 foot of ½-inch PVC to fit barb fitting
- Elbow to mate 1-inch PVC pipe to ½-inch Rubber O-ring (It must fit snugly inside the 1-inch pipe, but loosely over the ½-inch pipe; you may need to do some experimenting with different sizes.)
- Can of clear PVC cement
- Canning jar
- Brass barb fitting to match hose (It can be plastic but you will have to use epoxy instead of solder to attach it to the jar lid.)
- Petroleum jelly

TOOLS

- ☐ Drill
- ☐ Rubber bands
- ☐ Needle-nosed pliers
- ☐ Tin snips
- ☐ Vacuum gauge (optional)

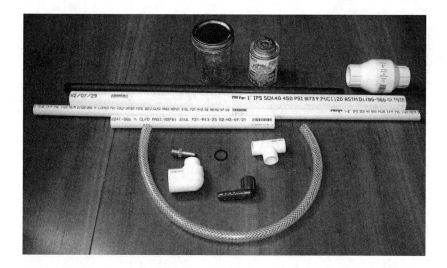

To see how the pump works, look at the drawing below.

The pump consists of a small diameter pipe (the piston) inserted into a larger diameter pipe (the cylinder).

The drawing shows the pump at two points during its operation. The top drawing shows air being drawn into the pipe by pulling out the piston. The bottom drawing shows air exiting the

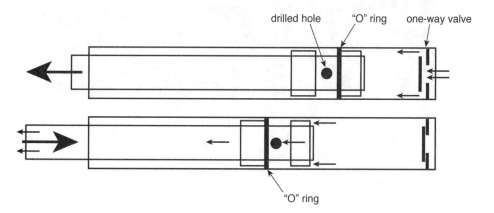

pipe when the piston is pushed back in. At the right end of the pipe is a one-way valve. This valve lets air into the pipe, but will not let air out of the pipe.

At the end of the piston is a homemade one-way valve.

The homemade one-way valve is made using a rubber O-ring that slides up or down on the piston. A hole is drilled into the piston, and two stops are attached, so the O-ring cannot slide past them.

When the piston is being pushed into the cylinder, the O-ring slides back, letting the air in the cylinder exit through the drilled hole and out through the hollow piston.

When the piston is pulled out of the cylinder, the O-ring slides toward the stop at the end of the piston and blocks the air from getting to the drilled hole. The right end of the piston is closed off, which seals the pipe. The left end is open to let the air escape.

This will all become easier to understand as you see the parts go together in the photos that follow.

The first step is to make the piston. Start by adding the stops that keep the O-ring from sliding. These stops are made from short pieces of the thin-walled ¾-inch tubing. Cut slots along the length of the tubing to allow it to be compressed onto the ½-inch piston tube.

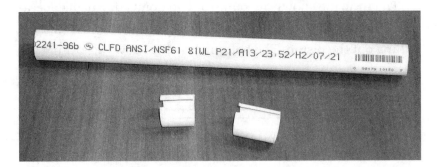

Using liberal amounts of PVC cement, glue the stops onto the piston, compressing them into place with rubber bands. The stops should be about ¾ inch apart.

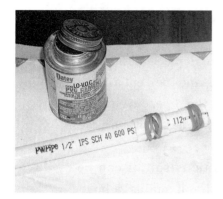

In the center of the space between the stops, drill three holes in the ½-inch pipe, spaced at roughly equal distances around the pipe, as shown below.

The commercial one-way valve has been cemented onto the 1-inch cylinder pipe in the bottom photo. Make sure that the arrow on the valve is pointing toward the cylinder.

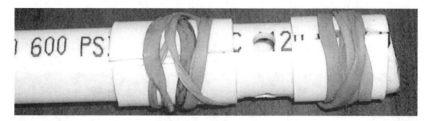

A short length of 1-inch pipe is cemented into the other end of the valve, and two reducing fittings are cemented to that so that it ends up with a hose barb at the end. (I found it convenient to use a right-angle reducing fitting, followed by a gray threaded pipe and a gray hose barb that threads onto it.) Be sure to cement any threaded parts, to ensure a vacuum-tight seal.

Next, cement a ½-inch T-fitting onto the far left end of the piston pipe to act as a convenient handle.

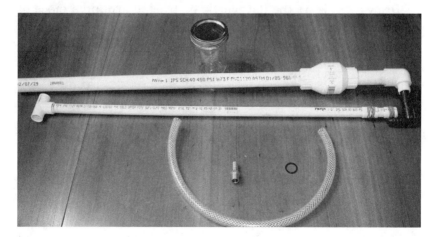

Now make a thin plastic disk to close off the right end of the piston pipe. Make it by flattening a small piece of the thin-walled ¾-inch pipe, and then cut a circle out of it. To flatten the pipe, cut a slot in it lengthwise, and then heat it over a stove to soften the plastic.

When the plastic is soft, it can be flattened by pressing it against a heat-resistant surface, such as a tile countertop or concrete floor. Use a pair of pliers to hold the tube. Place the bottom of a drinking glass on the plastic to flatten it and cool it at the same time.

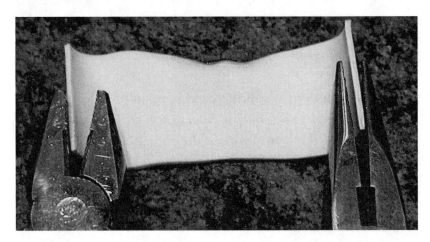

Draw a circle on the flat plastic by tracing around a piece of ½-inch pipe.

Cut the circle out of the plastic using tin snips or some other strong cutting tool.

Glue the disk onto the end of the piston pipe. Be gener-ous with the glue. Allow the glue to dry completely, and then remove the rubber bands.

Next, slip the O-ring over

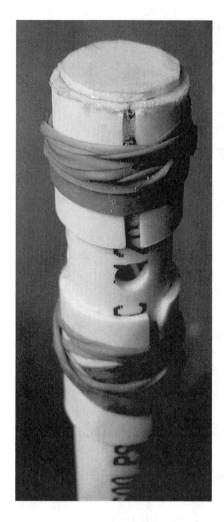

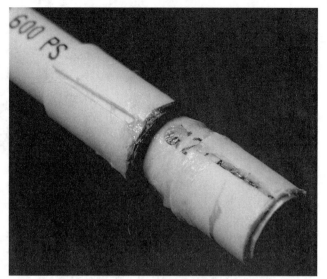

the stop at the end of the piston, and into the space between the stops, where the holes were drilled. Add a generous amount of petroleum jelly to lubricate the O-ring and ensure a vacuum-tight seal.

The piston can now be placed into the cylinder. It should be somewhat of a tight fit getting the O-ring into the cylinder, but the petroleum jelly should help lubricate it as it slides in.

The hose can now be attached to the hose barb. To make this easier, place the end of the hose into a cup of very hot water for a few minutes to soften it. It will now fit easily onto the barb, and shrink to a good, tight fit.

The completed pump is shown on the following page, attached to a vacuum gauge from an automotive parts store. The gauge is showing 23 inches of mercury, although with more pumping that figure can increase to 26 inches.

Finally, make a bell jar, a jar that will be connected to the pump using the hose. You will then be able to put things into the

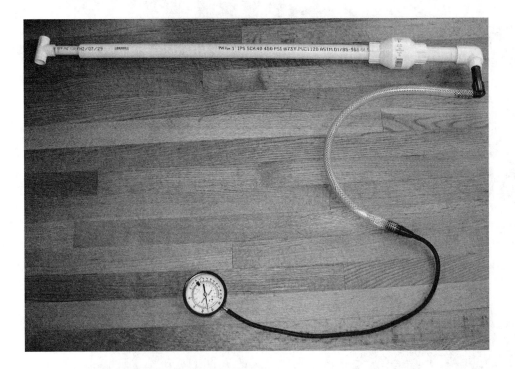

jar, and then pump the air out of the jar. Canning jars work best, since they are made to withstand a vacuum.

To make the bell jar, drill a hole in the center of the jar lid and attach a hose barb. In the photo you can see a brass hose barb soldered into the hole. You can use a plastic hose barb if you like, and use epoxy to attach it to the jar lid.

The photo on the following page also shows a pair of large marshmallows placed in the jar before screwing the top on. When the air is pumped out of the jar, the marshmallows will grow to twice their normal size. Then they grad-

ually shrink, as the air leaks out of the marshmallow and into the evacuated jar. When you later remove the hose from the jar, air rushes into the jar and the marshmallows shrink into dense, wrinkled candy with the consistency of a gumdrop. They feel heavy compared to normal marshmallows, although they weigh the same.

Combination Vacuum and Pressure Pump

The next pump is actually simpler to build (but more expensive because it uses two factory-made one-way valves).

This pump can draw a vacuum like the first, but it can also inflate beach toys and air mattresses, or it can pump water. Whether pumping water or air, the stuff goes in one end and out the other, instead of traveling through a hollow piston like the first pump.

To make it a little less expensive, use ¾-inch valves instead of 1-inch valves.

SHOPPING LIST

- ➲ 4-foot length of ¾-inch PVC pipe
- ➲ 2 3-inch lengths of ¾-inch PVC pipe
- ➲ ¾-inch PVC T-fitting
- ➲ 2 ¾-inch PVC one-way valves (also called check valves)
- ➲ 4-foot wooden dowel that fits inside the pipe
- ➲ Rubber stopper whose large end fits snugly into the pipe
- ➲ 3-inch-long wood screw
- ➲ Petroleum jelly

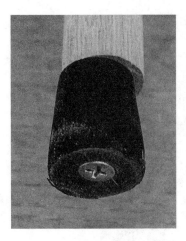

Start by drilling a hole in the end of the wooden dowel just a bit smaller than the wood screw. Also drill a hole in the rubber stopper. Next screw the rubber stopper onto the end of the dowel, as shown above.

You can control the thickness of the rubber stopper by tightening the screw to make it fatter if necessary. The stopper must completely block the pipe, yet still be able to slide in and out when lubricated with petroleum jelly.

The next step is to glue the PVC pieces together. Make sure that the arrows on both one-way valves are going in the same direction!

Here is a close-up of the one-way valve before gluing. On this valve, the arrow is molded into the plastic to the left of the label.

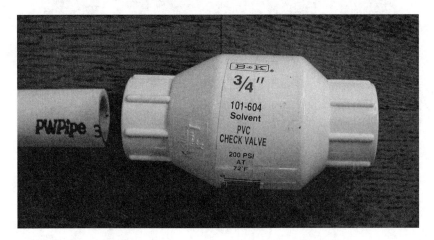

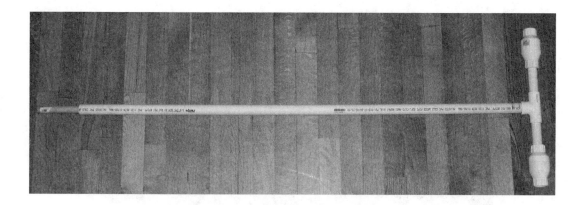

When the glue has dried, liberally grease the stopper with petroleum jelly, and slide it into the pipe. As you slide the stopper in, you will hear air coming out of the valve whose arrow points away from the pump. As you then pull the stopper back out, you will hear air entering through the other valve.

This pump is very effective for pumping water. The inside diameter of the pipe is 0.8 inches. With each 36-inch stroke of the piston, the amount of water pumped is:

$$36 \, \pi \, (0.4)^2 = 18 \text{ cubic inches}$$

That's about 10 fluid ounces, or 0.3 liters.

Propeller Toy

This classic toy was well known before Leonardo da Vinci was a boy, and it may have influenced some of his aerodynamic ideas. There are also stories about Orville and Wilbur Wright playing with propeller toys as kids.

This gizmo is easy to make. It's nothing more than a propeller on a stick, but the physics behind its stability in flight is not so simple.

SHOPPING LIST

- Block of soft pine approximately 8 inches long, 2 inches wide, and ½ inch thick
- 10-inch dowel, ¼ inch in diameter
- White glue

TOOLS

- Drill or auger with a ¼-inch bit
- Wood file or shaping tool, or a whittling knife (Power tools such as a drum sander or belt sander make the job go much faster.)

Start by drilling a ¼-inch-wide hole through an 8-inch block of soft pine.

Next, remove the wood from the corners of the block. If you are using a knife, hold the block in your left hand, and shave away the wood on the right side of the block. To make the propeller shape, remove only the wood on the top right side. The left side will remain untouched, and the right side is shaved down to a sharp edge. See photo on page 42.

Now turn the block over, and repeat, shaving off the right side only, so the propeller blade is a thin piece of wood at a pronounced angle to the hole.

Now hold the wood block by the blade you have just made, and carve the other end of the propeller in just the same way as the first. Again, only the right side is shaved down to the bottom; the left side is unshaved.

A knife, while traditional, is not the fastest, easiest, or safest way to remove the wood. Using a wood file, shaving tool, or planer is better. A power sander will do the job even faster.

The wood can be left in its rough whittled form, or it can be sanded smooth. You can paint the blades, or draw designs on them with felt-tipped markers.

Now glue the dowel into the hole. The photo shows a dowel that is 9½ inches long. The dowel can be a little shorter or a little longer, but a shorter dowel will make a less stable flight, and a longer dowel adds unneeded weight. The optimum length is something you will want to experiment with, as each hand-carved toy will be slightly different.

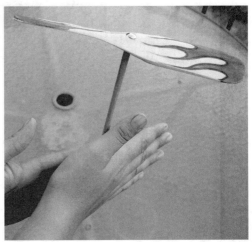

How to Fly It

Hold the dowel against your left palm using your right fingertips.

Quickly slide your right hand forward and your left hand back, so your left fingertips are against your right palm. The propeller toy will fly up and land a short distance away.

The photo below shows some toys made by hand in Africa. I support the artisans there by offering these hand-painted toys on my Web site (www.scitoys.com).

Rocket Engine

Rockets delight children and adults alike. The arc of smoke, the idea of flight, and the fun of action at a distance combine to make rockets a wonderful toy. To these pleasures this project will add the forbidden glee of playing with matches, and the satisfaction of learning how something actually works. Unlike most of the rest of this book, this really *is* rocket science.

The rockets you are about to build are perhaps the smallest toys in this book. They travel about 10 feet, which is nonetheless 150 times their length. They are hot to the touch and can blister a hand that catches them, so *they must only be fired outdoors on a fireproof surface,* such as a driveway (with your garage door closed). Since they are notoriously inaccurate, the center of the driveway is suggested to keep them from landing in dry grass. As always when playing with things that burn, have water—a bucketful or a garden hose—nearby in case of accidents.

SHOPPING LIST
- Aluminum foil
- Book of paper matches (or a box of wooden matches)
- Paper clip
- Straight pin

TOOLS
- Scissors
- Safety goggles

Start by cutting a narrow strip of aluminum foil, about 1 inch wide and 3 inches long. Tear the head off of a paper match (or cut it off with the scissors), leaving as little of the paper as possible attached to the match head. Set the match head on the foil near one end, with the top of the head facing down toward you.

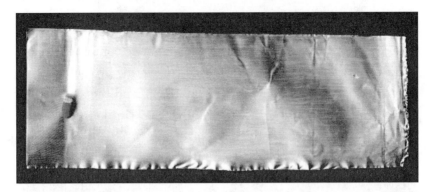

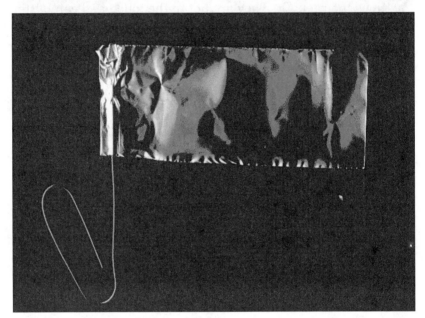

Roll the edge of the foil over the match head, forming a tube with the match head firmly in the center. Unfold the paper clip and insert one end of the wire into the tube so it touches the top of the match head (not the paper). Now press the foil tube flat to hold everything in place.

Fold the foil over one more time (or twice if you like), and tear off the excess foil. Now twist the foil at the free end tightly (without tearing it). Keeping the wire touching the match head at all

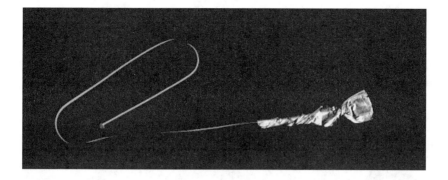

times; twist the end with the paper clip tightly around the paper clip.

Cut off the excess foil from the top of the rocket (the end away from the paper clip). You are now finished with the rocket, which means you're ready to build the launcher.

The rocket launcher is made from a straight pin stuck through a piece of cardboard (such as an empty matchbook) or a wad of aluminum foil. The pin should point upward at a 45-degree angle. Remove the rocket from the paper clip and slip it over the pin so that the pin is now where the paper clip used to be.

To launch the rocket, first aim it in a safe direction, where it

won't start a fire or melt anything it hits. *Put on a pair of safety goggles.* Then light a match, and hold it under the rocket.

The heat of the flame will cause the match head to catch fire inside the foil. The hot gases from the match head will have only one way to escape: through the hole where the pin is. As the gases go one way, the rocket goes the other, with a sharp hiss and a trail of smoke.

Depending on how big the match head was, and how much foil was used, the rocket will travel anywhere from a few inches to 20 feet.

If your rocket did not fly, but instead burst through the side of the foil, you will need to build one with more foil (by rolling it one more time before tearing off the excess). If it hissed but didn't leave the pin, check to make sure that your next rocket slides easily up and down the pin. The pin should not block the tube completely, but should leave some space between it and the foil. In other words, the pin should be skinnier than the paper clip. Be careful not to squash the tube flat while handling the rocket; instead, hold the rocket gently by the end containing the match head.

WHY DOES IT DO THAT?

A match head makes good rocket fuel because it carries both the burnable material and the oxygen needed to burn it. The burnable material in this case is the sulfur in the match head. The oxygen comes from compounds like potassium chlorate and potassium nitrate that are mixed with sulfur.

When the match head is heated, the sulfur and the potassium chlorate combine to make sulfur dioxide gas and potassium chloride. This generates a lot of heat, which heats up the gases.

The hot gases are under a lot of pressure, and their molecules hit the walls of the rocket at great speed. Where the nozzle of the rocket lets the gases out, there is no wall to hit, so there are no molecules pushing in that direction. The difference in pressure between the nozzle end (no pressure) and the opposite end (high pressure) is what pushes the rocket forward.

3

ELECTRICITY AND MAGNETISM

Ring Launcher

This very simple gizmo keeps amazing me with new tricks. My friend Jef Raskin first described it to me, and we have both been having fun with variations ever since.

All it consists of is 10 little magnetic beads on a carbon graphite composite rod. The beads all repel one another and arrange themselves in a way that demonstrates the effect of gravity—the beads at the bottom are closer together than the ones at the top, forming a clear mathematical progression.

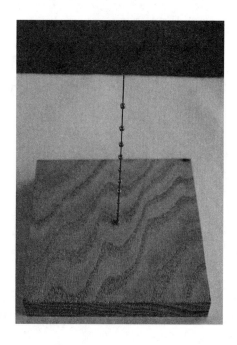

The fun doesn't stop there, though. If you turn the rod upside down the beads will flow down to the other end and arrange themselves, once again, in the same pattern, always staying separate.

Push down on the top bead, and then let it go. The ring shoots up into the air and sticks to whatever ferromagnetic surface it hits, such as a filing cabinet or refrigerator. The effect is nonlinear; adding one more bead makes the top bead go farther than adding the previous bead did.

There are two configurations you'll want to play with. In one, a bead is glued to each end of the rod, with eight (or more) beads in the middle. This form allows you to stick it to the refrigerator, and peel it off like a zipper.

If you let it stick to the refrigerator, then pull the top or the bottom, the magnets bunch up at the other end. Then peel it off, and they jump back one by one into the original sequence.

It's mesmerizing just to turn it upside down again and again, and watch the beads orient themselves, flowing like water down the rod.

The second form is the Ring Launcher. In this form, the rod is glued into a hole in a block of wood, so it stays upright (or at a 45-degree angle if you want maximum horizontal range).

You can launch the rings at steel targets, or have competitions or battles between two launchers. Or you can glue a bead to the top and just play with it on your desk, shooting the beads up to the top, where they are silently repelled back down by the top bead.

Building the Ring Launcher

SHOPPING LIST

- 10 magnetic beads (available at www.scitoys.com)
- 10-inch graphite composite rod (10 inches long, 0.03 inches in diameter)
- Super glue
- Wooden block

TOOLS
☐ Drill

You probably don't need any further explanation, but for the sake of completeness, the construction steps are shown below.

You start with a wooden block. Drill a small hole in the center.

The rod for your Ring Launcher is made from the same high-strength graphite composite as is used for high-end tennis racquets, fishing poles, and bicycles. It will bend easily and spring back to its original form. If bent too far, it will snap cleanly in two pieces.

The beads have holes in them that just fit over the rod. They are the same high-strength magnets as others you will find at www.scitoys.com. Like the others, they are gold plated to keep out moisture and withstand corrosion and abrasion.

Keeping the rod upright, glue it into the hole. Super Glue works well. Other glues take longer to harden and don't hold the slippery rod as well.

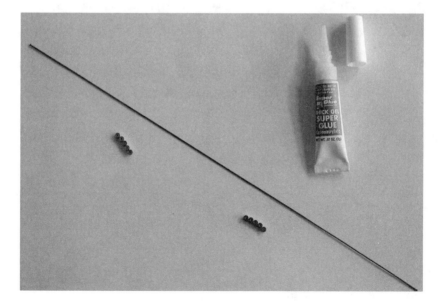

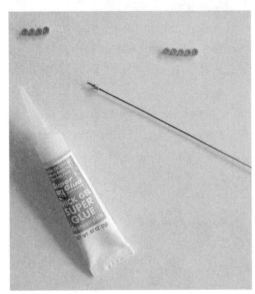

Place one bead at the bottom of the rod, against the wooden base, and let the glue stick it there.

Now place the rest of the beads on the rod one by one, making sure they repel one another. If they stick, turn the last one you placed on the rod over.

That's it! Now you have created a Ring Launcher.

To make a Zip Rod, don't use the wooden block.

Glue one bead at the end of the rod.

Let the glue harden, then place the remaining beads on the rod, so they repel.

Glue the last bead to the far end of the rod, again making sure it repels the bead before it. There should be a good deal of space between the last two beads, so you can watch the beads move and stack themselves.

Rail Gun

A rail gun is a device for accelerating an object by running electric current through it along a pair of rails. When large amounts of power are used, the rail gun becomes a potent weapon. But the principles can be demonstrated safely by using a smaller amount of power, such as from a 9-volt battery.

If you use a small amount of power, you can reuse the rails many times. If you use larger amounts of power, the rail gun becomes a one-shot device, as the rails are destroyed in the shooting process due to arcing and flexing of the rails.

There are several types of rail guns, each with a different method of accelerating the object. This version is called a linear homopolar motor.

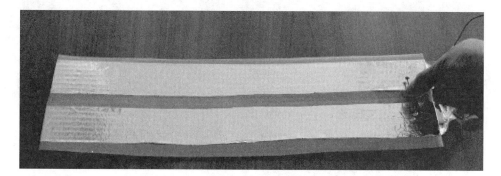

SHOPPING LIST

- ⊃ Cardboard or wood for a base (This can be any length; the one pictured is about 18 inches long and about 6 inches wide.)
- ⊃ 2 strips of aluminum foil, 2 inches wide, and 2 inches longer than the base
- ⊃ 2-inch length of steel wire, such as from a coat hanger
- ⊃ 2 disc magnets, plated in a good conductor, such as gold (available from www.scitoys.com)
- ⊃ White glue
- ⊃ 9-volt battery
- ⊃ 2 alligator clip leads

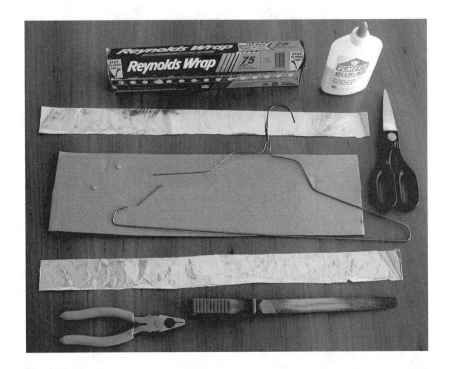

Start by spreading a thin layer of glue on the base, to hold down the foil strips.

Adhere the two foil strips onto the base, about ½ inch apart, and smooth them down with your fingers to remove wrinkles, as shown. The extra 2 inches are left hanging over one end to make it easy to clip on the leads for the battery.

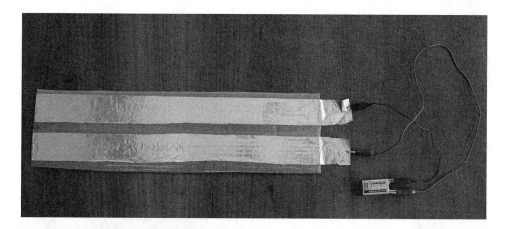

Next, attach the battery with the clip leads. Don't worry about battery polarity at this time—if the gun shoots the wrong way, you can reverse the battery.

File the ends of the coat hanger wire axle flat. This will allow the magnets to stick flat to the ends of the wire axle. If you file carefully and make the flat ends perpendicular to the wire, then there will be less wobbling as the magnets travel down the rails.

Now place the magnets on either end of the wire axle. The magnets should have their poles facing in opposite directions. The magnets repel one another when aligned this way, but they will still stick firmly to the steel axle. No glue is necessary, as the magnets are very strong.

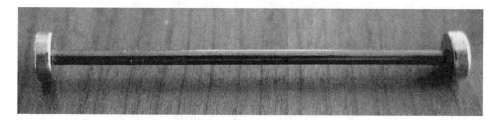

Firing the Rail Gun

To fire the rail gun, just drop the wheels on the rails. They will start to accelerate immediately. If they don't move at all, the magnets are probably not pointing the same way. Lift the wheels, flip one magnet, and try again. If it still doesn't work, check the battery and the connections.

If the wheels move in the wrong direction, you can either start them at the other end, or reverse the battery.

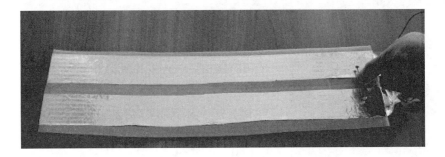

🤖 WHY DOES IT DO THAT? 🤖

The homopolar motor was one of the first motors ever built. Michael Faraday built one. There are many different designs. You can see one variation, shown below, on my Web site.

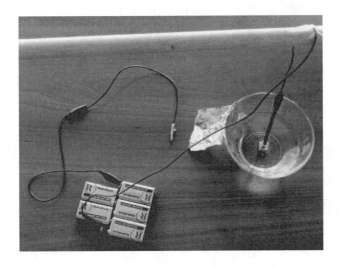

A magnet is placed with the north pole facing up in a bowl of vinegar. The magnet is incredibly powerful. Suspended over it is a heavy piece of copper tube or wire, hanging from a flexible stranded wire alligator clip. The vinegar covers the tube to a depth of about ½ inch. The tube does not quite touch the magnet but is free to swing around.

A piece of aluminum foil also rests in the vinegar. This can be a copper wire instead; it is not critical.

When the foil and the tube are connected to a source of about 30 to 50 volts (some 9-volt batteries connected in series) the tube starts to revolve around the magnet. You will also get lots of bubbles of hydrogen and oxygen, but, for this project, that's just a side effect.

In the experiment you created a homopolar motor. Unlike our previous motors, this one does not change the poles of an electromagnet from north to south and back.

As the current flows through the copper wire (or tube in our case), a magnetic field is created around it. This magnetic field interacts with the magnetic field of the magnet at the bottom of the bowl. The arrangement is set up so that the magnetic field in the wire exerts its force at a right angle to the magnetic field of the bottom magnet. This makes the wire circle that magnet.

Another simple homopolar motor is shown below.

The photo above shows the parts before assembly. There are three magnets—two small disks, and one cube whose north pole is marked with an *N*. There is a D-cell battery, and a bare copper wire formed into a shape that will make sense shortly.

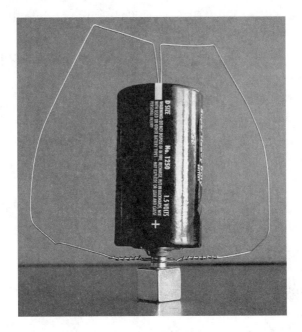

The two small disk magnets are set on top of the north pole of the cube magnet. The loop in the bottom of the wire form is then placed over the top of the two disks. The loop is just barely larger than these magnetic disks. The positive pole of the battery is then carefully placed on top of the disks, and the pointed end of the wire form is placed in the dimple at the top of the battery.

When the whole structure is balanced on the cube magnet, the wire begins spinning around the battery.

What is going on here is similar to what is going on in the vinegar version.

The current from the battery is flowing through the wire on both sides of the wire form. This creates a magnetic field around the wire. This field interacts with the field from the cube magnet and the disks. The wire's field creates a force at right angles to the field from the magnets. This causes the right side of the wire to be pushed toward you, and the left side to be pushed away from you. This action continues until the battery runs down.

Suppose you held onto the wire, and let the magnet move. You can do that by simply setting the battery on the table, on its side. You can see an animation of what happens at www.scitoys.com.

The wire rotates until it hits the table. The force continues after the wire can no longer rotate, and so instead of the wire rotating, the battery and the magnets rotate. This makes the battery roll along the table like a steamroller.

You can now make out the rail gun design. The rail gun is really just two of these battery rollers connected end-to-end. The battery is now external, and the wires have been replaced by the rails. The magnetic field in the rails creates a force at right angles to the magnetic field of the rail gun trolley's wheels, which causes them to roll.

4

COMPUTERS AND ELECTRONICS

Computer-Controlled Radio Transmitter

How would you like to send text messages to your friends without wires, an Internet connection, or monthly fees?

In this project you will build a very simple radio transmitter that you attach to a serial port on your computer. The computer then runs a free program that converts the words you type into radio signals that can be decoded by another computer, using a cheap radio receiver and a sound card.

With a little practice, you won't even need the second computer; the radio signals are in Morse code, which anyone can learn to decode in their head with a little practice. It also comes in handy as a secret language—or as a way to send long-distance messages with a pocket mirror.

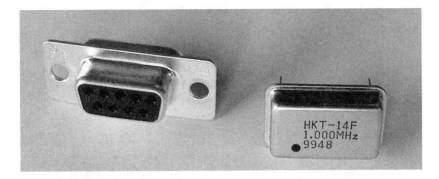

SHOPPING LIST

- 1-megahertz oscillator (You can use other frequencies if you have a radio that can receive them.)
- 9-pin serial port connector (You can take apart an old serial cable, or buy a new connector from an electronics or computer store: RS #232.)
- Insulated wire for an antenna, the longer the better
- Alligator test lead

For the first transmitter, you will connect the parts with alligator clips. This lets you quickly change frequencies by replacing the

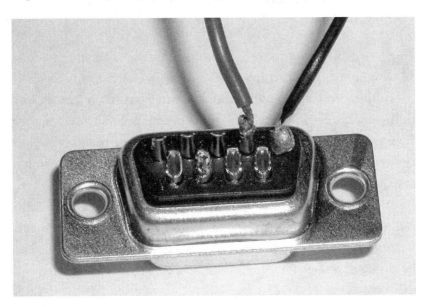

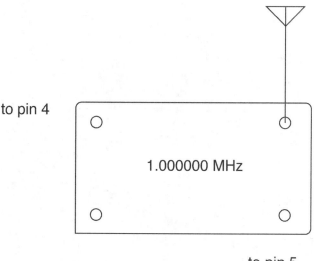

to pin 4

1.000000 MHz

to pin 5

1-megahertz oscillator with an oscillator that has a different frequency. Later you will build a version made with a socket for the oscillator, a printed circuit board, and a light-emitting diode that flashes Morse code along with the oscillator.

The first step is to cut the test lead in half. In these photos there are two different test leads to make it easier to see where the connections go. But unless you are making two transmitters (is your friend going to want to send messages back?), you can just use one. Remove a little insulation from the cut ends of the wire, and solder one of the cut ends to pin 5 and the other to pin 4.

Pin 5 of the serial port connector (the right-side wire in the photo on page 64) connects to the ground pin of the oscillator. Pin 4 of the serial port connector goes to the power pin of the oscillator. The drawing above shows the transmitter from the top (pins pointing down). The photo on page 66 shows the oscillator upside down, with the pins facing up.

The third alligator clip attaches to the antenna, which can be any long wire. It is attached to the output pin of the oscillator. The remaining pin of the oscillator (the one nearest the sharp corner) is not used.

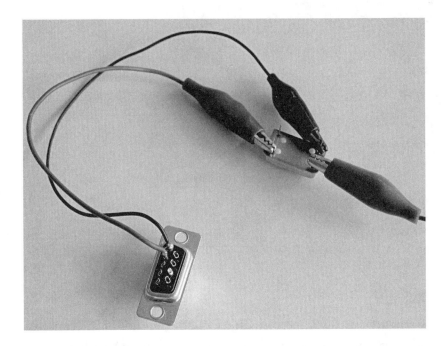

Your computer-controlled transmitter is now complete!

Controlling the Transmitter

To send a message, you need a computer program that can convert what you type into Morse code, and turn the oscillator on and off in the short and long pulses (dots and dashes) that are required.

A program to do that (for the Windows operating system) can be downloaded free from www.scitoys.com. The download link is http://scitoys.com/scitoys/scitoys/computers/radio/MorseCodeExe.zip

Save the ZIP file on your computer, use a ZIP file decompressor to unpack it, and then double-click on the resulting MorseCode.exe to start running it.

Once the program is running, you will see a window like the one on the following page. Type something in the window (such as "Hello there!") and then select Transmit in the Radio menu. Your transmitter is now sending your message.

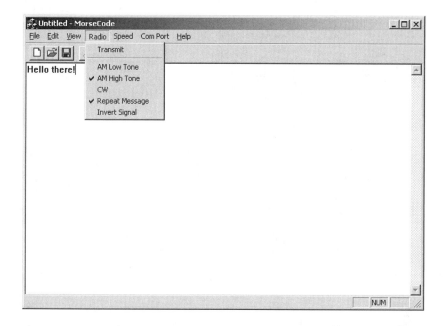

To receive the message, it helps to also select Repeat Message in the menu (as in the screen shot shown above). This will make the transmitter send the message over and over again, so you can more easily hunt for the signal on an AM radio dial.

Tune an AM radio to 1,000 kilohertz. If your radio has a numeric tuning indicator, this is easy. If the radio only has a dial with a few numbers on it, you will have to hunt around, tuning it until you hear clear Morse code coming from the speaker. It helps at this point to have the AM radio close to the transmitter's antenna.

You can select how fast the message is sent by using the Speed menu.

You can control which serial port to use through the Com Port menu.

The Radio menu has three selections that have not been discussed yet. The AM Low Tone selection sets the tone you hear in the AM radio to 500 hertz. The AM High Tone selection sets the tone to 1,000 hertz. The CW (continuous wave) selection is only for short-wave radios that have an SSB or CW mode. This selection does not modulate the radio signal, so an AM radio will just

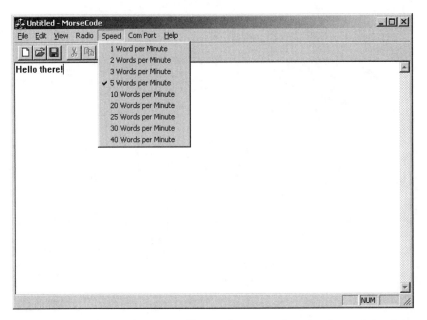

hear clicks. This selection allows the signal to be heard farther
away, but it requires a more expensive shortwave receiver. If you
are willing to spend some money, you can use the Grundig YB
400PE radio with great success. It usually sells for about $150.

If you are a computer programmer and would like to look at
the source code for this program, you can download it from the

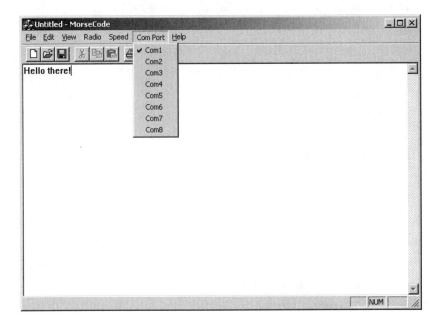

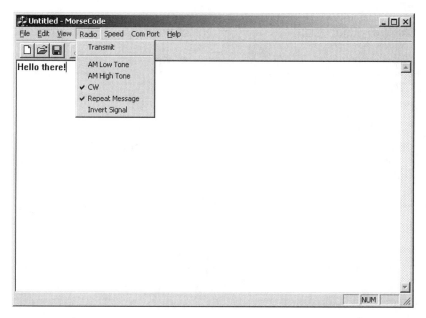

Web site. There is a much simpler command-line version of the program there as well.

Receiving the Code with a Computer

Until you have learned to decipher Morse code in your head, you will want to have a computer do it for you. There are many free programs floating around the Web that will do this. One such program, called CwGet, can be downloaded from http://scitoys.com/scitoys/scitoys/computers/radio/cwget.zip

I won't go into how it works (since I didn't write it), but it has a Help menu, and it is fairly straightforward to use. You will need an audio cable to connect the radio's earphone jack to the computer's sound card input jack, but that is all the hardware required.

You can see it working in the screen shot on the following page, decoding an endless loop of "hello there."

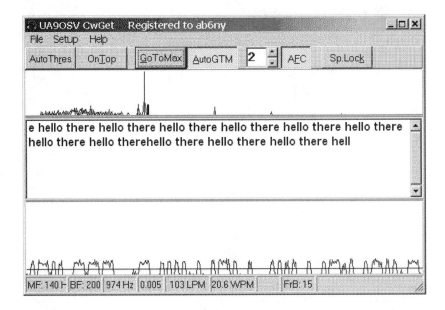

WHY DOES IT DO THAT?

Most of the important concepts for this project have been covered in other sections of this chapter.

The computer provides power to the oscillator through the DTR pin of the serial port. The program turns the DTR signal on and off, which causes the oscillator to turn on and off in return.

To make the signal audible in a cheap AM radio, the computer turns the power to the oscillator on and off 1,000 times per second while sending the dots and dashes of the code, and leaves it off in between the dots or dashes. This modulates the radio signal at a frequency your ears can hear. In AM Low Tone the audio frequency is 500 times per second.

In the CW mode, the computer does not modulate the radio signal. It just turns on the oscillator long enough for the dot or dash to be sent. In this case, the receiver does the work of converting the signal into an audible tone by using a circuit called a beat frequency oscillator. Your shortwave radio may have a switch labeled BFO, SSB, or CW that allows this circuit to operate.

Some Nicer Packaging

The computer program turns on DTR and also another signal called RTS, while sending the dots and dashes. In the version of the transmitter shown below, I have mounted a 14-pin socket to a general-purpose circuit board, and plugged the oscillator into that. A blue light-emitting diode (LED) is connected to the RTS pin of the serial port connector (pin 7). The LED flashes Morse code along with the oscillator, making an eye-catching effect.

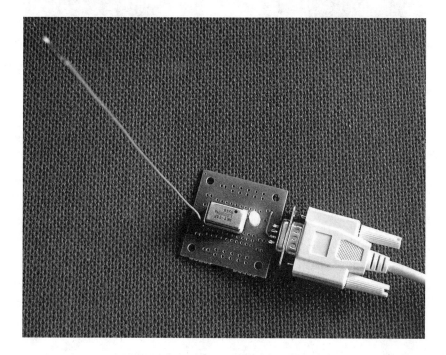

The serial port connector is wedged onto the printed circuit board by placing the board between the pins.

The wires that attach the serial port connector to the oscillator and the LED also serve to hold the connector onto the printed circuit board.

The antenna in this case is a 6-inch-long wire. In CW mode, this wire is all that is needed to receive the signal anywhere in the house. A longer antenna will allow the whole block to receive the signal.

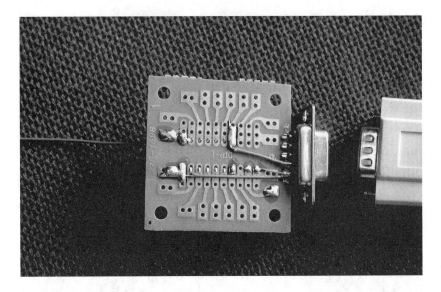

By replacing the 1-megahertz oscillator with a 28.322-mega-hertz oscillator, and connecting the transmitter to a large amateur radio antenna (10-meter beam), you could conceivably send signals from California to Texas. To do that, you would need to get an amateur radio license.

Learning Morse Code

There are many free programs to help you learn to decipher Morse code in your head. Some of them are: cw_play, MorseMad, NuMorse, and MorseCat. You can get them at www.scitoys.com, or you can just search for those names on Google.

Computer-Controlled Laser Data Transmitter

Although computers got their name from their ability to calculate, one of the main uses for computers today is in the field of communications.

Our modern telephone system is a large collection of computers, communicating with one another by means of laser light pulses through optical fibers.

You can do the same thing at home. In this project, you will build a laser transmitter that the computer will control, sending data by flashing the laser on and off. But you will eliminate the optical fibers and just send the light through the air, in what is called free space laser data transmission.

SHOPPING LIST

- Pocket laser pointer
- 9-pin serial port connector (RS #232) (You can take apart an old serial cable or buy a new connector from an electronics or computer store.)
- NPN transistor; almost any type will do, such as the 2N4401 or 2N2222A
- 470-ohm resistor with color codes yellow, purple, brown, and gold
- Light-emitting diode (A clear-lensed red LED is preferable, but almost any LED will do.)
- Generic printed circuit board (This is not really required, but it makes assembly easier: RS #276-159B.)
- Alligator test lead (This experiment used half of a red one and half of a black one, but a single test lead cut in half works nicely; it does not harm the laser to connect them incorrectly—just switch them around if the laser doesn't light up.)
- 9-volt battery clip
- Spring-type clothespin
- Screw or nail, about 2 inches long with a flat head
- Small block of wood
- 9-volt battery
- Electrical tape
- Glue

TOOLS

- Soldering iron

Modifications to the Laser Pointer

You will not actually modify the laser pointer, so it will be easy to undo the project and still have a working laser pointer. But you will be removing the batteries, taping down the on switch, and

inserting a small screw where the batteries were, to make it easy to connect the laser to the transmitter circuit you will build.

With the batteries removed, you can look into the back end of the laser and see the small spring that normally connects to the negative terminal of the battery.

You can also see the switch that turns the laser on—it is the little black box with the red button.

To make it easier to connect the little spring to your circuit, you will wrap some electrical tape around a small screw, and place the screw head against the spring. The tape will be wound

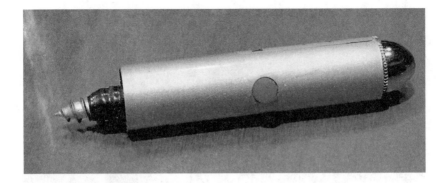

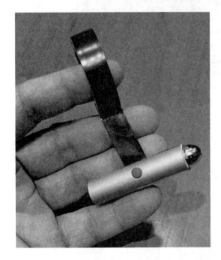

around the screw until it makes a snug fit inside the laser pointer housing, compressing the screw a little bit.

Next, use some tape to hold the on button down. You will be turning the laser on and off with the circuit, so the button will no longer be used; however, it must remain in the on position at all times.

The computer will communicate with the circuit through its serial port.

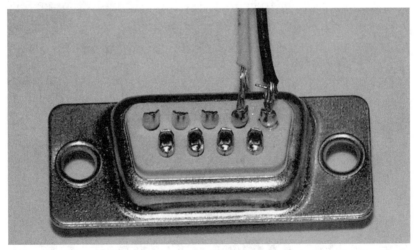

If your computer does not have a serial port, you can buy an inexpensive USB serial port to connect to your computer. It will work fine for this project.

You will use a 9-pin female serial connector, attaching wires to pins 4 and 5 only. Those pins are the Data Terminal Ready pin (pin 4) and the Ground pin (pin 5).

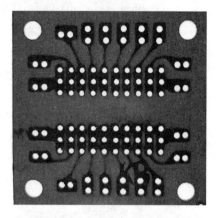

You will use a generic printed circuit board for this project, although all of the parts could simply be soldered together without it, or even connected with alligator test leads. But soldering the parts to a printed circuit board makes the project sturdy, and this guarantees the parts will stay connected.

One side of the board has copper foil printed on it. The other side of the board will have its components. The two sides of the board are called the *solder side* (where you do the soldering) and the *component side* (where the transistor, LED, and resistor will be).

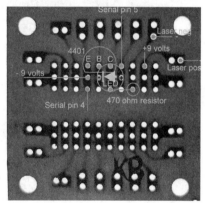

When you hold the board up to the light, you can see the shadow of the copper foil showing through on the component side.

The outlines of the components are drawn on the picture of the circuit board shown above. Some of the components seem to overlap in the drawing—this is because the parts will either be bent out of the way a little bit, or the leads will be left long, so one part will be higher than the others. You can choose to use either method of fitting the parts onto the board.

Solder the transistor onto the board first. With the flat side of the transistor facing the bottom of the page, the emitter, base, and collector wires will fit into three holes, and the wires will be soldered to the copper on the other side of the board.

Next, solder the LED onto the board by placing its leads through the proper holes and soldering them to the foil on the back side of the board. The LED has a flat side, which should face the left. The left lead is the cathode, and the right lead is the anode. The cathode will connect to the base lead (the middle lead) of the transistor, one hole down. The cathode lead is shorter than the anode lead.

Now solder the resistor in place. It will stand up straight, and the top lead will be bent over to go into the hole below the transistor's collector lead.

Cut off the excess leads on the solder side of the board, so it is neat and the leads don't accidentally bend onto one another.

Next, solder the battery clip wires onto the board, as shown in the photo on the following page. To prevent the heavy battery from pulling the wires off the board if it falls, feed the wires through a large hole and tie a knot in them before you solder them to the board. The negative (black) wire connects to the emitter lead of the transistor, one hole down. The red positive wire goes into the next unused hole to the right, just past the resistor.

Now solder the wires from the nine-pin serial connector to the board. The wire from pin 4 goes in the hole below the black battery wire, so it connects to the emitter of the transistor and the negative terminal of the battery. The wire from pin 5 goes into the hole just above the anode lead from the LED, so it is connected to the LED anode.

Cut the alligator test lead in half, and solder one half to the

hole that leads to the red battery wire, and the other half to the hole that leads to the bottom lead of the resistor.

In schematic form, the circuit looks pretty simple, since there are only four parts:

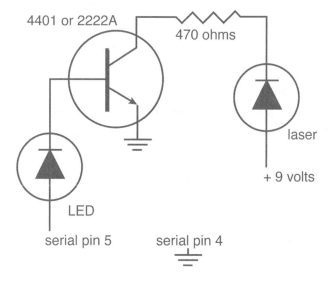

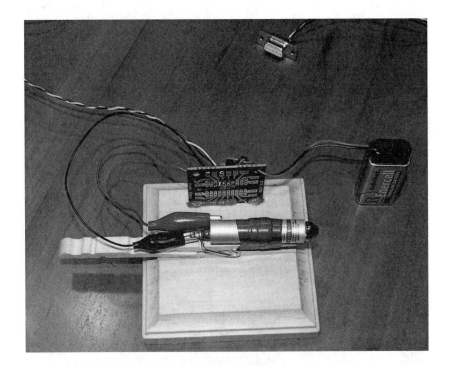

You are now ready to connect all the parts. Tape the laser to the clothespin as shown in the photo, and glue the clothespin to the block of wood. Insert the wedge-shaped end of half of another clothespin between the jaws of the first clothespin. This arrangement makes it easy to make very small adjustments to the vertical angle of the laser beam, which in turn makes it much easier to aim the laser.

Connect the positive alligator test lead to the barrel of the laser, and the negative test lead to the pointed end of the screw. Connect the battery to the battery clip.

Plug the serial connector into the serial port of the computer. The LED and the laser should both light up as soon as you do this. If they don't, carefully check all your connections to make sure that you didn't accidentally bridge two of the copper foil traces together with solder.

If the LED lights up but the laser does not, check the connections, and also check that the button on the laser is firmly depressed.

Controlling the Transmitter

To send a message, you now use the same computer program that you used in the Computer-Controlled Radio Transmitter project to convert what you type into Morse code, and turn the laser on and off in dots and dashes.

There is one small difference in the setup, however. The laser is not directly powered by the serial port, as the radio transmitter was. The laser circuit has a transistor switch in it. The transistor *inverts* the signal. This means that when the serial port is turned on, the transistor turns the laser off. When the serial port turns off, the transistor turns the laser on. This is caused by the simple circuit used. You could have used two transistors to prevent the inversion, but instead, you can simply tell the computer program to invert the signals before it sends them. This is easier and cheaper than adding another transistor.

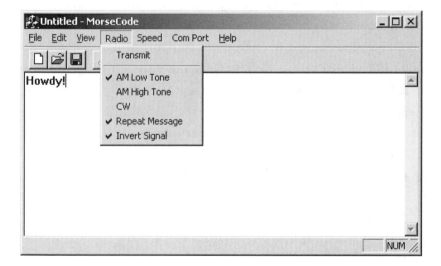

To receive the Morse code signals, you can use a very simple receiver, made from a piezoelectric earphone and a small solar cell.

You can also connect a phono plug to the solar cell instead of the earphone, plug it into the sound card of another computer,

and use the same receiving program as in the Computer-Controlled Radio Transmitter project.

🤖 WHY DOES IT DO THAT? 🤖

Each of the four components in the circuit performs its own special task.

The signal coming in from the serial port swings between 25 volts DC to −25 volts DC. The LED not only lights up to show that the signal got there but, because it is a diode, it cuts off the negative swings of the signal so the transistor only sees a signal of 0 volts or 25 volts. (Most serial ports never actually go much

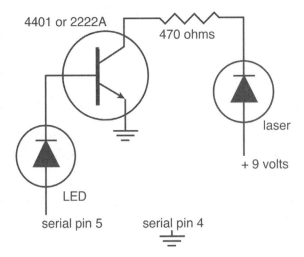

higher than 12 volts, but the RS-232 specifications allow as much as 25 volts.)

The transistor is set up to act as a simple switch. It has three leads—the base, the emitter, and the collector. The emitter is connected to the negative side of the power supply, which is called *ground* because it is often connected to the earth in electronics and radio circuits. The base is connected to the LED that lights up when the serial port signal is on.

When the base of the transistor sees the voltage go from 0 to anything above 1 volt, the transistor goes from the off state to the on state, allowing electrons to flow from the emitter to the collector.

The collector is connected to the 470-ohm resistor. This resistor is needed to prevent too much current from flowing into the laser, which would damage it.

You use a 470-ohm resistor because you want the current going through the laser to stay below 30 milliamperes. There is a simple rule for calculating how much resistance you need to limit the current. It is called Ohm's law. It says that the current is equal to the voltage divided by the resistance:

$$amperes = volts \div ohms$$

You have 9 volts and 470 ohms, so $9 \div 470$ is about 0.019 amperes, or 19 milliamperes (a milliampere is $\frac{1}{1000}$ of an ampere). This is enough to light the laser brightly, and yet well below the 30-milliampere limit that would damage the laser.

Finally, when the electrons flow from the emitter to the collector and then through the resistor, they reach the laser and cause it to light up, which sends a beam of light you can detect from as much as a mile away at night, or across the street in the daylight.

Fun with
Solderless Breadboards

When building a circuit for the first time, it is useful to have a way to change connections or parts placement quickly.

In the early days of electronics, temporary circuits were sometimes built on a piece of wood, similar to a breadboard. Building a first prototype came to be known as *breadboarding*.

Back when components like tubes and transformers were large and long wires were common, it was easy to solder and unsolder connections. Modern circuits, made with small transis-

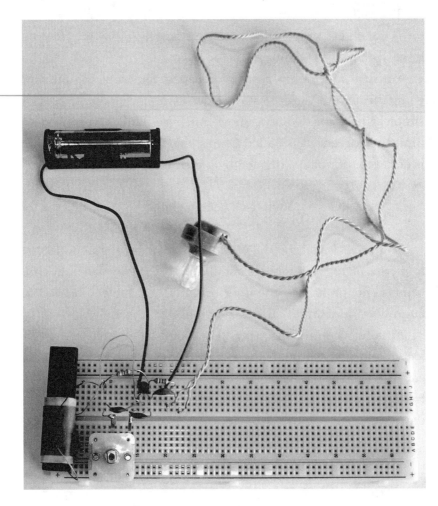

tors or many-legged integrated circuits, are much harder to solder, and they are especially difficult to *un*solder.

To make life easier, solderless breadboards were invented. These are blocks of plastic with holes into which wires can be inserted. The holes are connected electrically, so that wires stuck in the connected holes are also connected electrically.

The connected holes are arranged in rows, in groups of five, so that up to five parts can be quickly connected just by plugging their leads into connected holes in the breadboard. When you want to rearrange a circuit, just pull the wire or part out of the hole, and move it or replace it.

The photo opposite shows a complete radio, with an antenna coil, a tuning capacitor, a three-legged integrated circuit, a battery, an earphone, two resistors, and three capacitors. (This radio is the Three Penny Radio kit from www.scitoys.com, which is usually soldered using three pennies to connect the various parts.)

SHOPPING LIST

- ⊃ Antenna coil (You can wind one by hand, or use a much smaller coil with a ferrite rod inside, available from www.scitoys.com.)
- ⊃ MK484-1 AM Radio Integrated Circuit
- ⊃ Piezoelectric earphone
- ⊃ Tuning capacitor, from 0 to 160 microfarads
- ⊃ 100,000-ohm resistor (Its four colored bands will be brown, black, yellow, and gold.)
- ⊃ 1,000-ohm resistor (Its four colored bands will be brown, black, red, and gold.)
- ⊃ 0.01-microfarad capacitor (marked something like ".01M" or "103")
- ⊃ 2 0.1-microfarad capacitors (marked something like ".1M" or "104")
- ⊃ 1.5-volt battery
- ⊃ 1.5-volt battery holder
- ⊃ Solderless breadboard
- ⊃ Printed circuit board (optional)

TOOLS
☐ Soldering iron

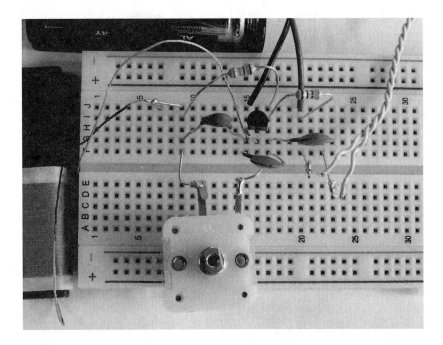

In the photo, you can see that the parts that are going to be electrically connected are plugged into one of the five holes marked A, B, C, D, or E, (or in this case, where the second half of the board was used, marked F, G, H, I, and J) in rows marked 1 through 63.

Along each side of the breadboard are two strips of power-supply rails, which makes it convenient to connect a battery when many parts need power. In this simple radio, only one part needs power, so you can plug the battery wires directly into the main circuit area.

The solderless breadboard is designed to accept the leads from parts such as resistors, integrated circuits, transistors, and other parts with round solid wire for leads. Some of the parts for the radio have thin, stranded wire that is not stiff enough to poke into the holes, or (like the variable capacitor) have flat strips of metal that are too big to fit into the holes. For these parts, solder

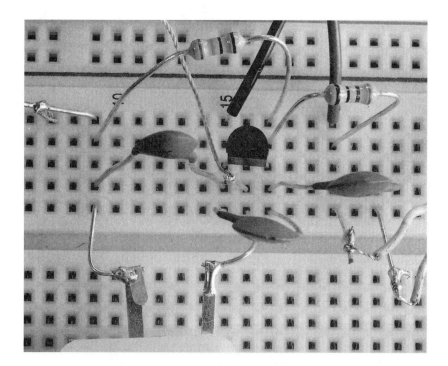

their leads to pieces of wire cut from the leads of other parts, such as resistors or capacitors. Most such parts have leads that are longer than you needed anyway, so they will fit more snugly onto the board with shorter leads. In the photo on page 88 you can see that the antenna coil, the variable capacitor, and the piezoelectric earphone have wires soldered to their leads to make it easy to plug them into the breadboard.

Having the holes arranged in a labeled grid is convenient for describing the parts layout. You can list each part and the letter and number of each lead:

* Antenna coil: J9 and G16
* Tuning capacitor: F9 and F16 (only the two rightmost leads are used)
* MK484 IC: H15, H16, and H17 (flat side facing row G)
* 100,000 ohm resistor (brown, black, yellow): I9 and J17
* 1,000-ohm resistor (brown, black, red): I17 and I20
* 0.01 microfarad capacitor (marked 103 or .01M): G9 and G15

* 0.1 microfarad capacitor (marked 104 or .1M): F15 and F17

* 0.1 microfarad capacitor (marked 104 or .1M): F17 and F22

* Piezoelectric earphone: F20 and F22

* Negative battery wire (black): J15

* Positive battery wire (red): J20

This makes it very simple to build the circuit and to double-check all of the connections.

Making the Circuit Permanent

Solderless breadboards are great for building circuits the first time, and getting them to work, or experimenting with design changes. But when you get the circuit working the way you want it to work, you will want to copy it to a more permanent form by soldering it onto a circuit board.

The printed circuit boards available at www.scitoys.com also have five holes that are electrically connected. The holes are

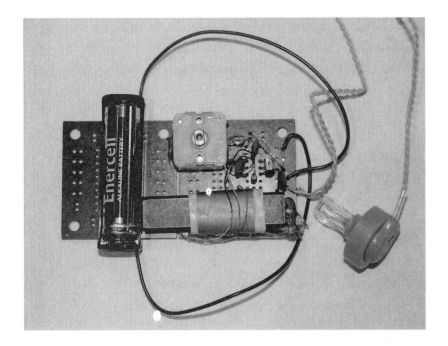

grouped into three holes and two larger holes, to make it convenient to connect larger wires leading out from the board, for
power connections and other external parts.

The radio shown below was built by a student as a first exer-
cise in soldering.

Here is the back side:

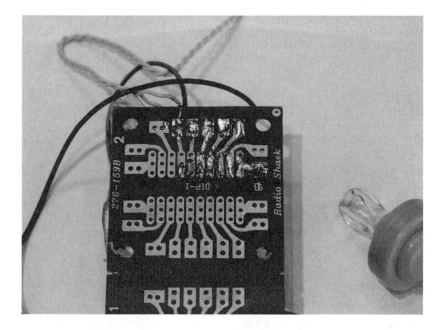

It worked the first time!

1-Watt Audio Amplifier

The solderless breadboard makes it easy to experiment with additions to the radio circuit. In this section, you will build a simple amplifier, so that a whole room can hear the radio through a speaker. The amplifier will not be ear shattering, but the output is impressive for a single transistor.

With the amplifier, the radio looks like the photo below.

Below is a close-up of the amplifier section:

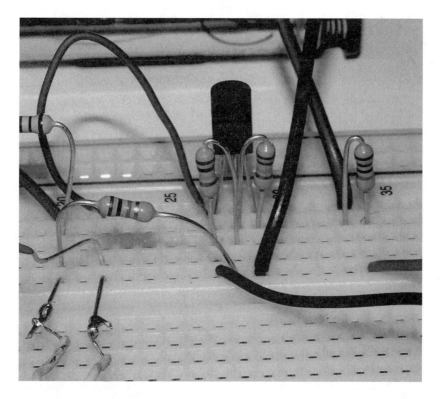

SHOPPING LIST

- ⊃ MPSW45A Darlington transistor
- ⊃ Small speaker
- ⊃ 2 100,000-ohm resistors (Their four colored bands will be brown, black, yellow, and gold.)
- ⊃ 10,000-ohm resistor (Its four colored bands will be brown, black, orange, and gold.)
- ⊃ 50-ohm resistor (Its four colored bands will be green, black, black, and gold.)
- ⊃ 9-volt battery clip
- ⊃ 9-volt battery

Using the labeled grid as before, the parts are connected this way:

* Jumper wire: J22 to I27
* 10,000-ohm resistor (brown, black, orange): G20 to F28
* 100,000-ohm resistor (brown, black, yellow): H27 to H28
* 100,000-ohm resistor (brown, black, yellow): I28 to I29
* MPSW45A: J27, J28, and J29
* 50-ohm resistor (green, black, black): I33 to I34
* Speaker: F29 to J33
* Negative 9-volt battery wire (black): F26
* Positive 9-volt battery wire (red): F34

A More Permanent Version

As before, you can copy the circuit onto a printed circuit board and solder all of the parts firmly in place.

WHY DOES IT DO THAT?

The heart of the amplifier is the transistor. You could have used a more ordinary NPN transistor, such as the 2N4401, but to get a louder sound, you use a special two-in-one type of transistor called a Darlington.

The Darlington transistor has two transistors in the same package, and can amplify signals much more than a single transistor can.

Transistors amplify a signal by acting like a variable resistor. You put the signal in at the base, and the signal controls how much current goes through the transistor from the emitter to the collector.

If you simply put the signal into the base, however, the transistor would turn off completely when the signal was low, and turn on completely when the signal was high. This is useful when you want to use the transistor as a switch, but you have to change the behavior to make a good audio amplifier.

When the signal is at 0, you want the output of the amplifier to be halfway between 0 and 9 volts (4.5 volts). You can arrange for this to happen by using a voltage divider. A voltage divider is two resistors, one connected to the positive side of the battery, and the other to the negative side. Where they meet in the middle, the voltage will be divided in half (if the resistors are the same).

Since current flows through the resistors all the time, you want their values to be high, so that not much current flows through them. This will prevent them from getting hot, and make the battery last longer. In your circuit you will use 100,000 ohms.

Large resistors in the voltage divider also make it easier for the signal to push the voltage higher or lower. This is a good thing, since it means our amplifier will be more sensitive. In this case, the signal from the radio is a little too strong, and the signal

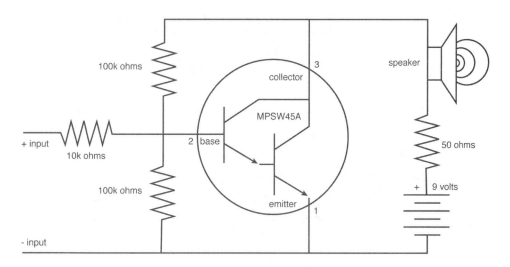

pushes the voltage too high and too low, causing distortion. So add another resistor, with 10,000 ohms, to match the signal to the amplifier.

The transistor can handle 1 watt before it gets too hot and its lifetime is reduced. If you let the full 9 volts get in, the circuit would draw over 2 watts, and while the sound would be nice and loud, the transistor would get quite hot and the battery would not last long. To make the amplifier draw only 1 watt, you put in a 50-ohm resistor to lower the current. You may have noticed from the photos that 100 ohms were used for the prototype; that was to keep the noise level down in the lab. You can think of this resistor as a volume control, although you can't adjust it without picking another resistor. A variable resistor that could handle 2 watts and go from 50 ohms to 150 ohms would let you vary the volume. You're the experimenter; go ahead and experiment.

Integrated Circuit Audio Amplifier

Like the simple amplifier in the previous section, this project starts with the solderless breadboard version of our Three Penny Radio, and adds an amplifier. But this amplifier has several important improvements over the simple one-transistor amplifier.

This project uses an integrated circuit with 10 transistors to amplify much better with much less power drain on the batteries than our simple amplifier.

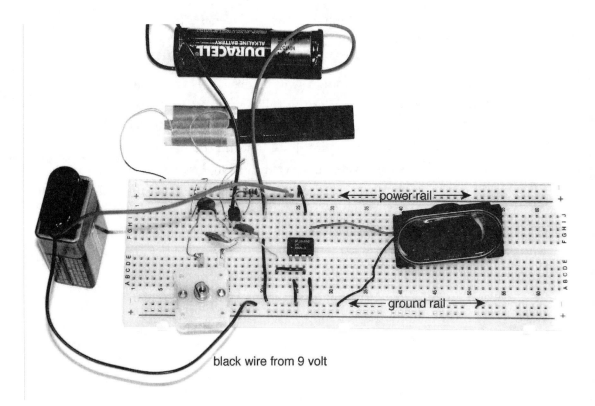

black wire from 9 volt

SHOPPING LIST

- ⊃ LM386 integrated circuit amplifier chip
- ⊃ Small speaker
- ⊃ Jumper wires
- ⊃ 9-volt battery clip
- ⊃ 9-volt battery
- ⊃ 10-microfarad capacitor
- ⊃ 220-microfarad capacitor
- ⊃ 0.033-microfarad capacitor
- ⊃ 0.047-microfarad capacitor
- ⊃ 10-ohm resistor (Its colored bands will be brown, black, black, and gold.)
- ⊃ 10,000-ohm resistor (Its colored bands will be brown, black, orange, and gold.)
- ⊃ 1,200-ohm resistor (Its colored bands will be brown, red, red, and gold.)

The Three Penny Radio normally has a piezoelectric earphone attached at points J-20 and E-22. You will replace the earphone with the amplifier.

On the following page is a close-up of the amplifier section.

Here you can see that the Three Penny Radio output at J-20 is connected to the ground rail below the blue line. This rail has all of its holes connected together. Connect the black (negative) wire from the battery to the ground rail. Connect the red (positive) wire from the battery to the power rail, just above the red line at the top of the photo. Having the power and ground connected to these rails makes it easier to connect the other parts, and makes it easier to see where all the connections are.

The other output from the Three Penny Radio is plugged into E-22.

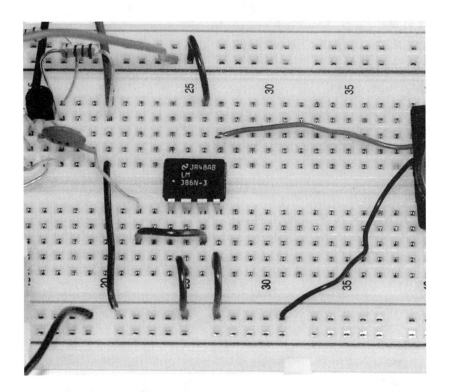

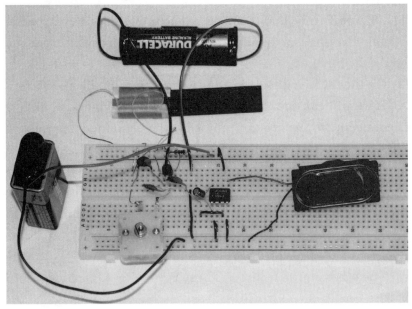

Using the labeled grid as before, the parts are connected this way:

* LM386 amplifier chip at E-24, E-25, E-26, E-27 and F-24, F-25, F-26, and F-27
* Jumper wire: F-20 to ground rail
* Jumper wire: C-22 to C-26
* Jumper wire: A-25 to ground rail
* Jumper wire: A-27 to ground rail
* Jumper wire: J-26 to power rail
* Speaker: red wire to H-27 and black wire to ground rail
* Negative 9-volt battery wire (black): ground rail
* Positive 9-volt battery wire (red): power rail

When all the wires are connected properly, you should be able to hear radio stations coming from the speaker. They will not be particularly loud, but you can make the volume 10 times louder with a simple adjustment.

Put a 10-microfarad capacitor connecting pins 1 and 8 of the integrated circuit (put the negative capacitor lead into hole D-24 and the positive capacitor lead into hole G-24). This bypasses a

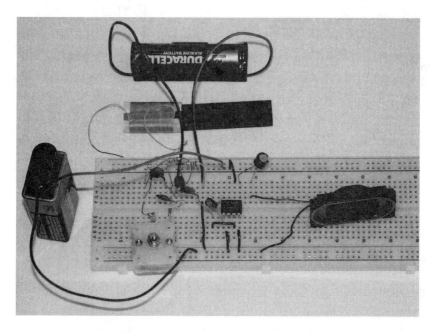

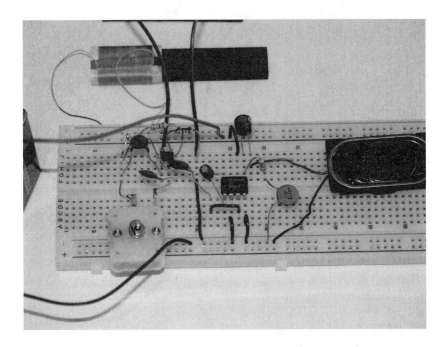

resistor inside the integrated circuit, boosting the gain from 20 to 200.

One problem with the circuit so far is that the speaker will get warm and the battery will not last long. This is because a certain amount of direct current is going through the speaker. Direct current does not make sounds, and so this current is a complete waste of battery power; all it does is heat up the speaker coil.

You can fix this problem by putting a 220-microfarad capacitor between the integrated circuit output pin (pin 5, in hole F-27) and the red speaker wire. Put the positive capacitor lead into hole J-27, and the negative capacitor lead into hole J-29. Move the red speaker wire to hole G-29.

To prevent the capacitor from changing the sound, you add two more components to create a filter that lets only audio frequencies get to the speaker.

Here you have moved the negative lead of the 220-microfarad capacitor to hole J-28, and you have moved the red wire of the speaker to H-28. Put a 10-ohm resistor into holes G-27 and G-30.

A 0.047-microfarad capacitor goes into hole F-30 and the ground rail.

The amplifier should be working pretty well now. But the speaker has a high, tinny sound due to its small size. It would be nice if it had a little more power on the lower frequencies, what is called bass response.

You can arrange to amplify the lower frequencies more than the high ones. Make another filter from a 0.033-microfarad capacitor and a 10,000-ohm resistor, and connect that between the output pin and pin 1 of the integrated circuit. This will feed some of the low frequencies back into the amplifier to be amplified again.

Put the 10,000-ohm resistor in holes C-24 and F-29. Put the 0.033-microfarad capacitor in holes I-27 and I-29.

Now your amplifier has a more pleasing, realistic sound.

A More Permanent Version

As before, you can copy the circuit onto a printed circuit board and solder all the parts firmly in place.

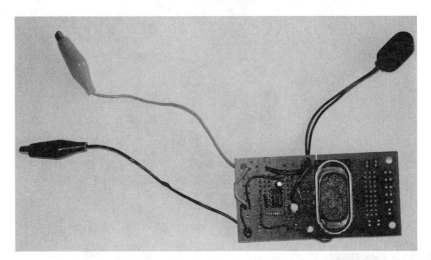

This amplifier can be used for many projects, from amplifying crystal radios to amplifying the signals from electric fish (see page 138).

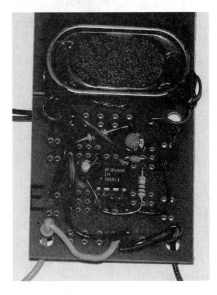

5

MATHEMATICS

Kaleidocycle

As this little paper sculpture turns inside out, it changes colors. First yellow, then blue, then red, then green, and then yellow again. It will cycle through the colors as long as you keep turning it inside out. Known as a Kaleidocycle, or Flexahedron, the toy was invented years ago by a bored mathematics student. It goes together quickly, and it can keep you happily fidgeting for hours. Best of all, it's free. Just print out this Web page on a color printer: http://scitoys.com/scitoys/scitoys/mathematics/paper_toy_5x.jpg

Print the picture out as large as the paper allows. Cut off all white parts of the paper, and follow the simple directions outlined below. If you don't have access to a color printer, you can copy the blank pattern on page 104 and use paints, markers, or crayons to color in the diamonds according to the labels on page 105.

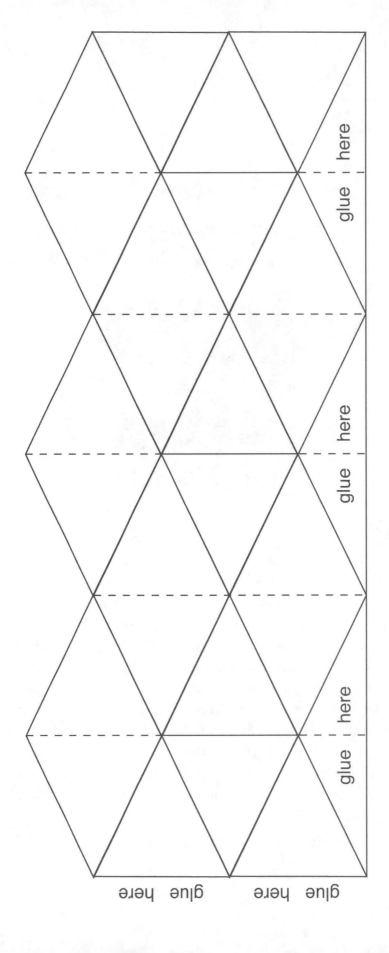

The printed picture from my Web site has two patterns on it. The pattern on the right has lots of extra lines and text on it, to help you fold your first toy properly. The one on the left will be the second toy you build, and it will look better because there is no extra printing on it. Once you are good at folding them, you can use: http://sci toys.com/scitoys/scitoys/mathematics/ paper_toy_4x.jpg. That version has two patterns without any extra printing.

After you have printed out or copied and colored your pattern page, cut out the colored pattern. Then fold it carefully along the lines that separate the colors, and along the lines that connect the points of the diamonds. Some of the folds will eventually go inward and some outward (if you're using the pattern in this book, inward folds are indicated by a dashed line, outward folds by a solid line), but in this step fold the paper back and forth both ways, and crease it well.

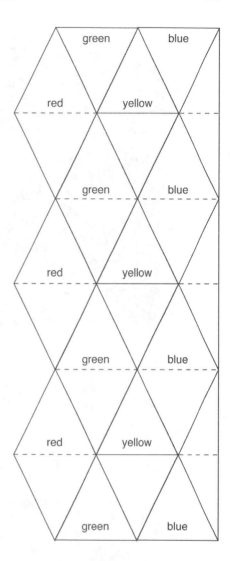

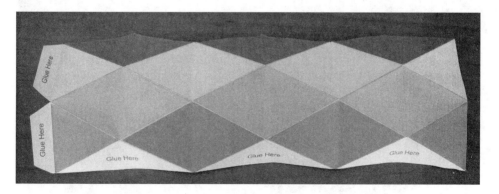

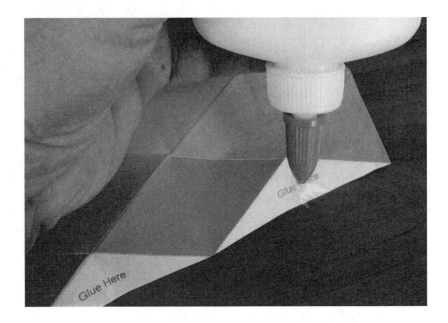

Now that the paper is nicely creased along the fold lines, it is time to spread some glue on the first little triangle that says "Glue Here" as shown above.

Use a toothpick or a small piece of paper or cardboard to spread the glue into a very thin film. Using too much glue will both make a mess and take a lot longer to dry.

Next, fold the paper so that the blue diamond above the blue half-diamond wraps under and fits onto the part of the paper you just covered with glue. Bend the blue diamond in half to make it easier to get the paper aligned properly.

Repeat this process with the other two blue diamonds, so you have a snake-like collection of six tetrahedrons, all hinged together.

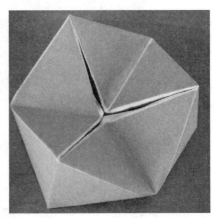

In the last step, spread glue on the remaining two tabs at the end, and carefully insert both of them into the other end of the snake, in the open end. Now pinch that end closed, so the tabs are glued to the paper, holding the snake's tail firmly in his mouth. It will form a ring.

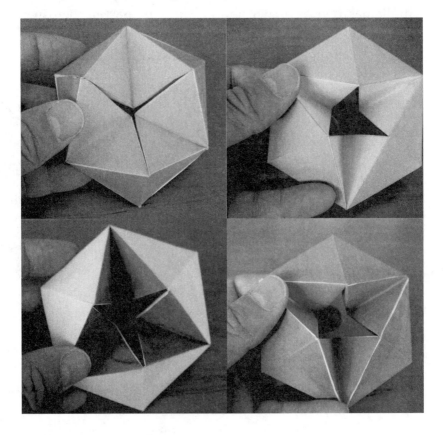

It is important now to let the glue dry completely. If you play with the toy too soon it will come undone.

When the glue is dry, you can start turning the ring inside out, gently pushing the center up from the bottom, and the outside parts down. The colors will change from one to the next as you keep playing with it.

Geodesic Dome

Some years ago I built a geodesic dome out of ½-inch galvanized steel electrical conduit, to serve as an aviary for chickens and small parrots. I wrote a computer program to calculate the proper lengths of steel tubing, and drew the diagram shown below.

The dome is made from three different lengths of tubing. I used colored stickers on the tubes to mark the different lengths— red for the long ones, violet for the medium lengths, and green for the short ones. You can see those colors in the drawing on the Web site.

I smashed the ends of the tubes flat with a hammer, and then drilled holes through the flat ends for a bolt to go through to connect up to six of the tubes together. Finally, I wrapped aviary wire around the dome to keep the birds in and predators out.

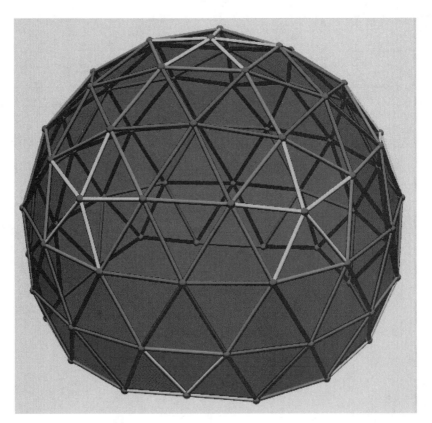

The completed dome is 18 feet high and more than 20 feet in diameter.

For this project, something a little more modest in size is required—something less than 3 feet in diameter, so it can fit through a door. This first dome can be built using bamboo kebab skewers and gumdrops.

SHOPPING LIST
- 90 skewers 7¾ inches long
- 85 skewers 7½ inches long
- 80 skewers 6½ inches long
- 11 green gumdrops
- 15 orange gumdrops
- 50 red gumdrops
- Cellophane tape

TOOLS
- Wire cutters

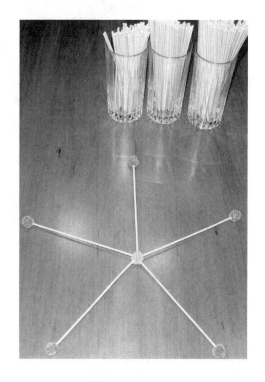

With the skewers all cut to the proper lengths (using a pair of wire cutters), the first step is to make the central pentagon of the dome. Insert five short skewers into a green gumdrop, and stick a red gumdrop onto each of the five ends of the skewers.

Next, stick five medium skewers into the red gumdrops. The pentagon will not be flat. Notice that by making the skewers different lengths, you can ensure that

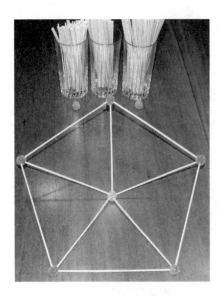

the proper three-dimensional form will take shape.

The dome is made of pentagons and hexagons. Each of the five sides of the pentagon has a hexagon attached to it whose six interior braces are made from the longest skewers. The sides of the pentagons and hexagons are all made from the medium skewers. Carefully study the labels on the photo below.

In the gumdrop dome there are green gumdrops at the center of the pentagons, orange gumdrops at the center of the hexagons, and red gumdrops everywhere else.

When five hexagons have been attached to the central pentagon, and five more green gumdrops connected at the bottom
between the hexagons, you will have a structure that looks like a
dome.

This is about as far as you can go with gumdrops alone. At
this stage, all of the weight of the dome is being held up by the
stickiness of the bottom gumdrops. But gumdrops can only hold
a small amount of weight before the skewers start to pull out.

To solve this problem, use cellophane tape to hold the skewers together. You will only need to do it for the red and orange
gumdrops, where you can easily connect the sticks in pairs, across
the gumdrops. Attach one end of a 4-inch length of tape to one
skewer, lay the tape over the gumdrop onto the opposite skewer,
and then fold the edges of the tape around the stick so the sticky
sides hold together. You can see this in the photo on the following page.

You will end up with a nice dome, but it won't hold up too well if you move it around a lot because the gumdrops are heavy and don't hold the skewers well enough. But you get the concept, and it was nicely color coded.

Another Dome

A second dome, more portable and permanent, can be made from plastic soda straws and small brads, the kind used for fastening paper.

SHOPPING LIST

- ⊃ 90 straws 7¾ inches long
- ⊃ 85 straws 7½ inches long
- ⊃ 80 straws 6½ inches long
- ⊃ 76 small brads

TOOLS

- ◻ Ice pick

First, cut all the straws to length.

Next, poke holes ¼ inch from both ends of each straw with a sharp knife or an ice pick. Be careful!

Next, make six five-sided stars (from the short straws), and fifteen six-sided stars (from the longest straws) by sticking the

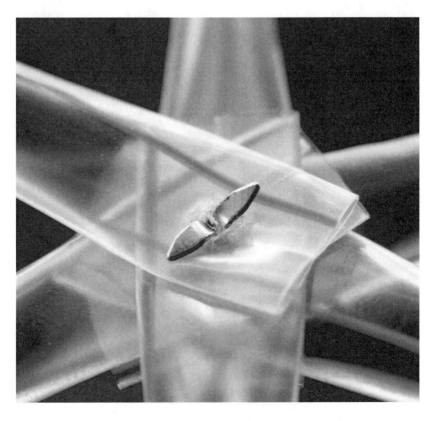

brad through the ends of five (or six) straws, and bending the ends of the brad over to secure them.

Begin assembling the dome by connecting one side of the center pentagon to one side of a hexagon. The photo above shows one hexagon for clarity, but you will be putting two hexagons on at a time, so there are always six straws meeting at each brad.

The photo also only shows one medium-length straw forming an edge, but you will be adding the edges on all sides, closing the brad whenever you have six straws connected.

In the photo on the bottom of the preceding page, five hexagons are connected to the central pentagon, and five more pentagons (without sides yet) connected between the hexagons.

The finished dome is 2 feet 2¾ inches tall (68 centimeters). It is lightweight, and can be tossed around without breaking. It will

fit through doors or in the back of a station wagon. As you can see, the geometry of the finished sculpture is quite beautiful.

My large dome is still housing chickens out in my backyard. It is amazingly strong—I had to climb all over it to wrap the aviary wire on it, and it never budged.

I put a small plastic shed (7 feet high, 6 feet long, and 8 feet wide) next to it, and removed three struts to form a door, and screwed the aviary wire to the shed, so the shed forms a nice door and airlock that won't allow the chickens to escape.

A large plastic tarp attached with bungee cords forms a waterproof roof. The small parrots live in the top half of the dome, and the chickens have the bottom half.

Paper Geodesic Dome

The same ideas can be used to build a dome made from triangular sheets of material instead of rods. The edges of the triangles have the same lengths as the rods did.

If you choose to use paper, you can print out a template on a printer, and then simply cut it out, fold it, and tape the edges together to get a dome.

You can print out the following page from the Web site, or make a photocopy of it, enlarging it if you want a larger dome.

If you want an even larger paper dome, you can cut the drawing into smaller pieces, so when your printer resizes them to fit the page, you will have larger triangles. Use tape to reconnect the smaller pieces into one large piece. In the drawings on pages 122 and 123, I have duplicated the central pentagon to make it easy to see how they fit together. You will want to cut one pentagon away before taping the pieces together.

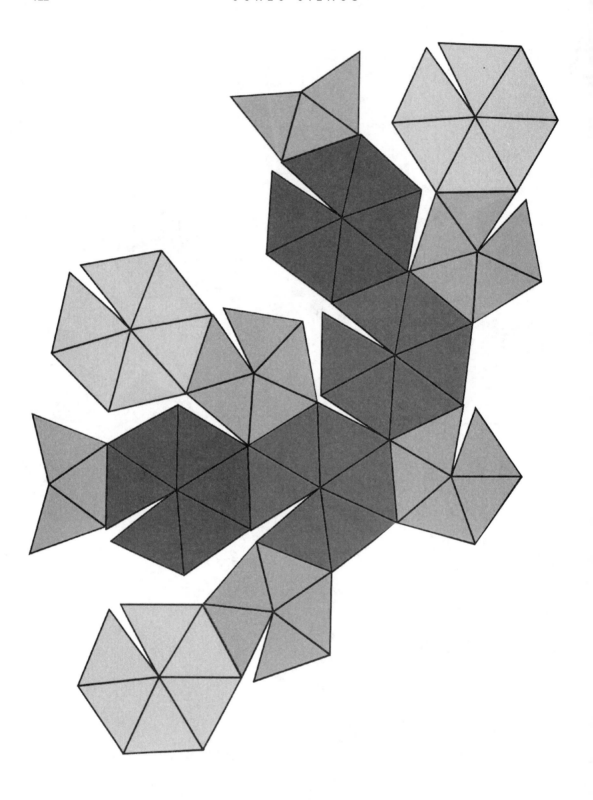

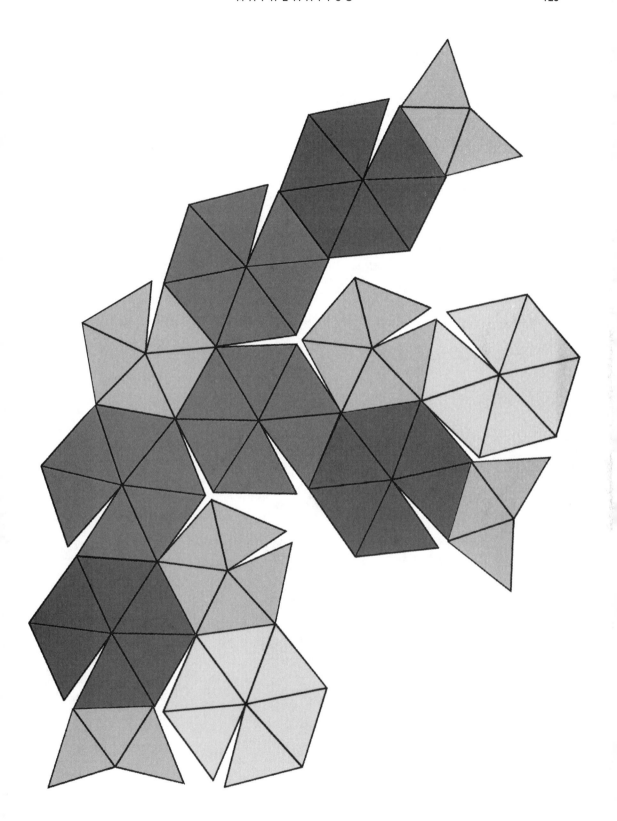

6

BIOLOGY

Photography Through a Microscope

Amateur Microscopes and Cheap Cameras

A simple toy microscope and a disposable camera can team up to make stunning photographs of very tiny things. The technique is simplicity itself. Just focus the microscope carefully as you would normally. Then place the camera lens right up against the eyepiece, and snap the picture.

The results are surprisingly good.

The photos on page 127 were taken with an $8 disposable camera. While the camera's optics are clearly not designed to

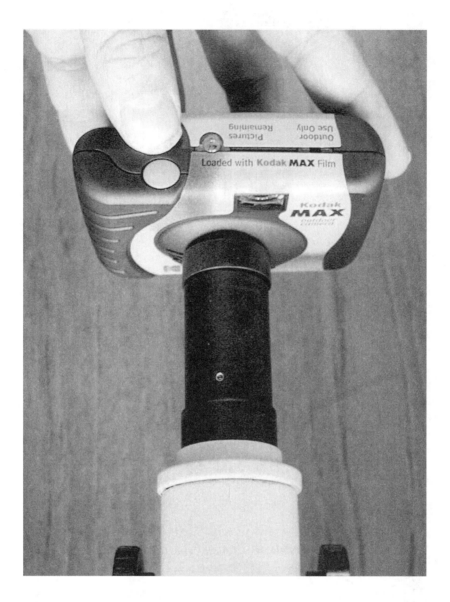

compete with those of expensive cameras, the images hold up to close examination.

The main drawback to the disposable camera is the fixed focus. Later you will see the results of using a camera that can adjust the focus to take sharper pictures. But the better camera costs 100 times as much as the disposable. (You might want to split the difference and get a medium-priced camera. After all, a $100 camera would probably take excellent pictures.)

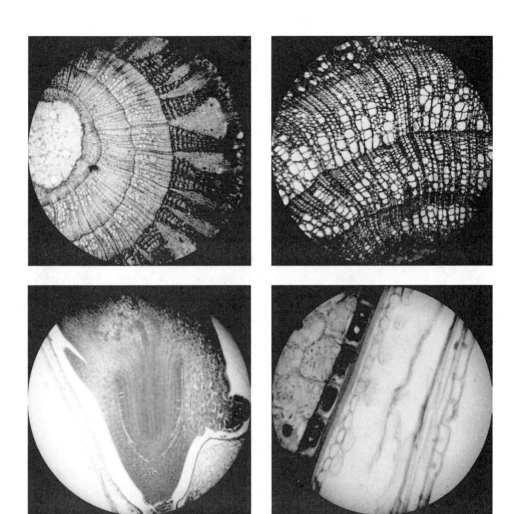

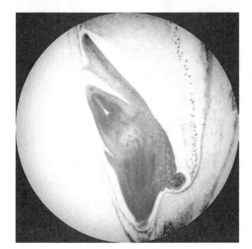

top left: Basswood (Tilia) stem, 4x objective.
top right: Basswood (Tilia) stem, 10x objective.
center left: Wheat kernel, 10x objective.
center right: Wheat kernel, 40x objective.
left: Wheat kernel, 4x objective.

The photos were developed normally and scanned using an inexpensive color scanner (about $40).

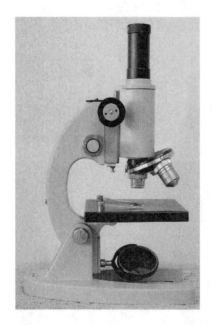

A Selection of Microscopes

The microscope shown in the photo at left is an inexpensive model commonly used by students and amateurs. It has no optics beneath the stage except a concave mirror. It has three objective lenses, and a single eyepiece, for magnifications of 40 times, 100 times, and 400 times. This one is made by a company called Labo; it costs about $100.

Better microscopes give better results. The next step up is a microscope with a *substage condenser.*

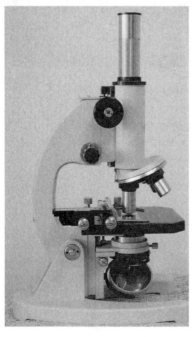

The microscope shown bottom left is made by Eagle; it costs about $300. It has several important features. Below the stage where you place the slide, there is a large lens system called a condenser. This system of lenses focuses the light onto the specimen.

A substage condenser makes the images sharper and allows more control of the illumination.

In addition to the condenser, this microscope has a mechanical stage. This is a device for holding the slide that allows the operator to turn two knobs to move the slide smoothly left and right or up and down.

This microscope has interchangeable

eyepieces, so the operator has a choice of a wide field of view or a close-up. This feature is not as important as the substage condenser, since a higher power eyepiece will not show more detail (to see more detail you would have to use a higher power objective lens). It is like blowing up a photograph. It makes it easier to see some parts of the image, but you don't get more information.

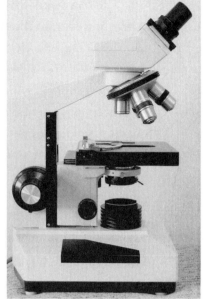

The next step up in amateur microscopes is the model shown at right. This is the Celestron Research Microscope, and it sells for about $700.

The Celestron adds binocular eyepieces, an illumination lamp, and an *oil immersion objective lens* that allows magnifications of 1,000 and 1,500.

Professional microscopes add features such as *phase contrast*, *differential interference contrast*, laser scanning confocal imaging systems, epi-fluorescence microscopy, and other features that greatly enhance the ability to see structures and features of very small things. Some manufacturers of these microscopes are Nikon and Carl Zeiss. In general, these microscopes have prices a little beyond the range of the average amateur.

Using Digital Cameras

Disposable cameras are nice because they are cheap, and they have small lenses that make it possible to simply aim them down the eyepiece and snap a picture.

More expensive cameras generally have larger diameter lenses, which makes things a little more complicated. The lenses must be shielded from light getting in around the eyepiece, and usually some sort of macro lens must be used to make the eyepiece fill the photo.

The more expensive single-lens reflex cameras have an advantage, however, in that they have through-the-lens viewfinders, so you can see exactly what the picture will look like before you click the shutter release.

The problem with film cameras though, is that to see your pictures, you must wait to have the film developed. With digital cameras, you can get immediate results. Many inexpensive digital cameras have small lenses that are a good match to the size of the eyepiece of the microscope, and they often have electronic viewfinders that show you exactly what the camera sees as you take the picture.

One digital camera I like to use for microphotography is the Camedia 2020 (or the more powerful 3030) from Olympus. The Camedia takes pictures of reasonable resolution (1,600 x 1,200 pixels for the 2020, and 2,048 x 1,536 pixels for the 3030). It also can take short movies, a process you will experiment with later.

Some Digital Photomicrographs

The photos below were taken with the Camedia 2020, in high-resolution mode (1,600 x 1,200 pixels). The pictures were cropped to eliminate much of the black border around the eyepiece.

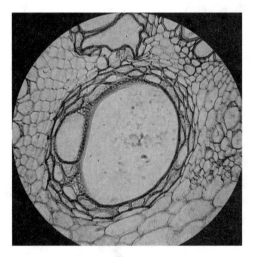

left: Pumpkin stem, sectioned and stained, 40x objective.

opposite, top left: Basswood (Tilia) stem, sectioned and stained, 4x objective.

opposite, top right: Same slide, 10x objective.

opposite, center left: Frog blood, 40x objective.

opposite, center right: Onion skin, 10x objective.

opposite, bottom left: Pumpkin stem, sectioned and stained, 4x objective.

opposite, bottom right: Pumpkin stem, sectioned and stained, 10x objective.

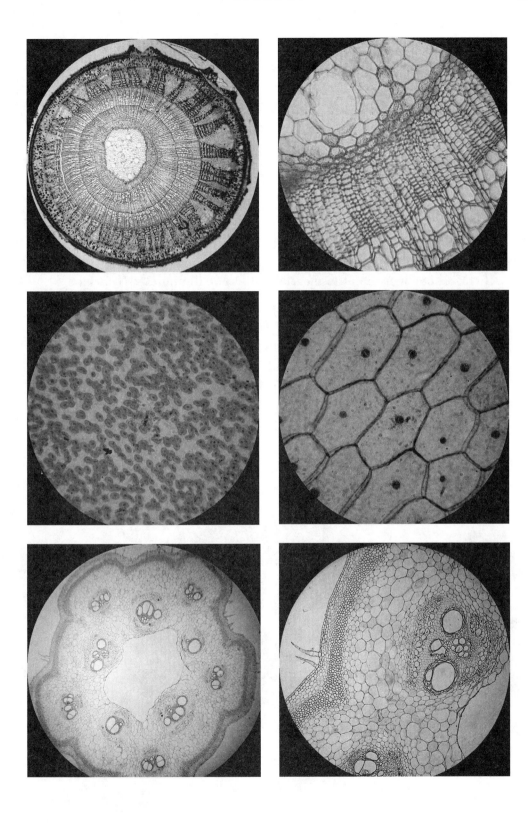

Comparing Microscopes

An amateur on a budget will want to know just what an extra few hundred dollars will buy in the way of better photos. Below are some comparison photos, showing the difference between the $100 Labo microscope and the $700 Celestron.

The Labo has no substage illuminator, so the slides were examined with the concave mirror reflecting a bright cloudy sky onto the slide.

The Celestron's illuminator was used, which caused the colors to be bluer than those of the same slide illuminated by bright clouds.

All photos were taken with the Camedia 2020 in high-resolution mode. Since the actual magnification is a function of the eyepiece, the camera lens, and the settings of your monitor, rather than give a guess at how much the image is magnified, I will instead give the power of the objective lens, which is what determines the resolving power, or the amount of detail that can be seen.

On the left side of the image below, the Celestron shows a little more detail, but the difference is actually quite small. For low mag-

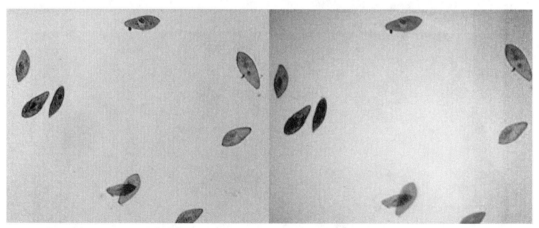

Paramecia stained, 4x objective.

nifications, the inexpensive microscope holds its own quite well.

At higher magnification, the Celestron (right) shows better contrast and sharper detail.

With the 40x objectives below, the value of the substage condenser begins to show. The Celestron shows more detail, and the contrast between the macronucleus and the rest of the cell is higher. However, the inexpensive microscope is still doing an outstanding job with these stained slides. The use of

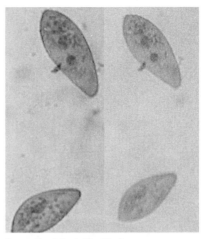

Paramecia stained, 10x objective.

stains to enhance contrast can make up for a lot of missing optical resolution.

The Celestron has a 100x oil immersion objective that the other microscopes lack. At this resolution you can start to see

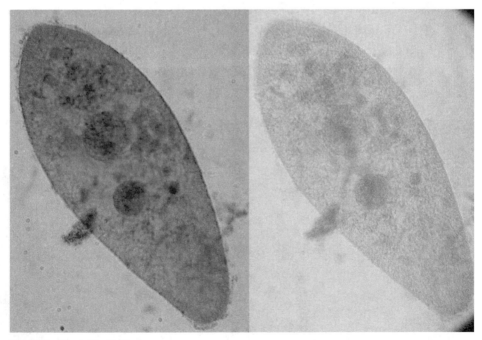

Paramecia stained, 40x objective.

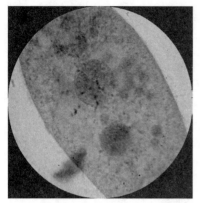

Paramecia stained, 100x oil immersion objective.

structure in the macronucleus, especially when focusing in and out through the depth of the structures.

Controlling Contrast

Most of the work in microscopy is aimed at increasing the contrast between the subject and the background. One of the earliest tricks (probably used by the first microscopist, Anton von Leeuwenhoek) is *dark field illumination*. You have used this trick to see small details without a microscope any time you have watched dust motes in a sunbeam. The dust motes are illuminated by bright sunlight, but the dark background is not. The contrast between the bright dust and the dark shadows allows you to see the dust that is normally invisible.

There are two main methods for achieving dark field illumination in a microscope. For low powers, where the distance between the objective lens and the subject is relatively large, you can simply aim a strong light down at the subject, and remove

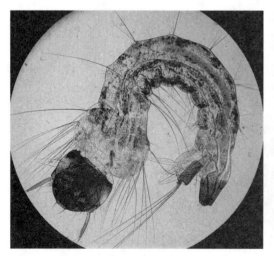

Mosquito larva, bright field illumination.

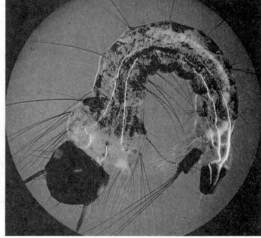

Mosquito larva, dark field illumination.

any light coming from below. For higher magnifications, where the objective lens is almost touching the subject, this method will not work. What is done instead is to make use of the substage condenser and a small black dot placed between the condenser and the source of light.

The condenser projects the image of the black dot onto the objective lens, so the background is dark. However, the light from around the edges of the black dot is focused on the subject, and fine details catch the light and are clearly visible against the dark background.

Video Through a Microscope

As I mentioned, the Camedia 2020 and other digital cameras can take videos as well as photographs. Below are a couple of examples.

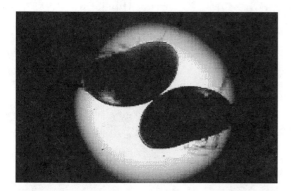

These subjects at top left are ostracods, tiny crustaceans found in freshwater ponds.

You can watch the videos at www.scitoys.com, but the still shots will have to do for now.

For more ambitious projects, nothing beats a DV camcorder. The video below was shot with a Sharp Digital Viewcam, and transferred to the computer via an IEEE 1398 (also known as Firewire) cable.

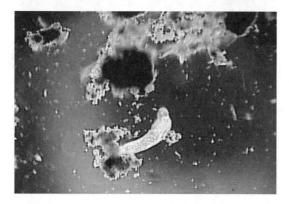

The subject at bottom left is a microscopic worm (possibly a planarian) and a lot of small ciliates such as paramecia.

Using a Video Camera as a Microscope

Some cheap computer video cameras have adjustable focus lenses that can focus almost right up to the lens. This allows you to aim the camera straight up, put a microscope slide on the lens ring, and focus on the slide.

Shown at right is a Logitech QuickCam camera, facing up, with the microscope slide on top of the lens ring. The video

shows the procedure, with a voiceover from the author (me).

The subject is once again an ostracod. (I have a bunch of them I've been keeping as pets in a big jar for almost five years now. They eat fish food.)

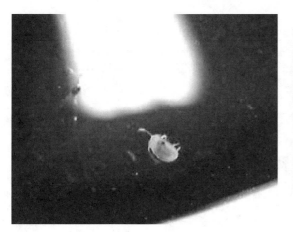

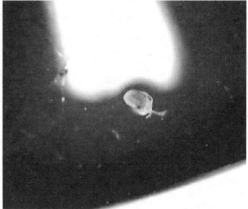

Listening to Electric Fish

At your local tropical fish store, you can find a particularly wonderful creature called an Elephant Nose fish, or more properly, *Gnathonemus petersii*, a member of the Mormyrid family.

Besides the obvious visual interest of the animal, there is another fascinating aspect to it—it emits pulses of electricity into the water, with which it locates food, other fish, and potential mates.

It is easy to listen in on these electrical signals with simple, inexpensive equipment, such as a piezoelectric earphone or a small amplifier.

SHOPPING LIST

- Elephant Nose fish (Gnathonemus petersii)
- Fish tank
- Piezoelectric earphone

A piezoelectric earphone is a simple device described earlier in this book. You can get one at Radio Shack in a crystal radio kit. It is very sensitive to small electrical signals, so it's perfect for detecting the electrical signals from the Elephant Nose fish and converting them into audible sounds.

To listen in on the fish, simply put one wire from the earphone into the water at one side of the fish tank and the other wire into the water at the other side of the fish tank, and put the earphone up to your ear.

At www.scitoys.com, you'll find a sound file, so you can hear what it sounds like. (Scroll down the home page until you find Biology—Listening to Electric Fish on the menu on the left side.)

If the fish is still, the electrical pulses will be infrequent. But as the fish moves around, it increases the frequency of the pulses until it is almost a buzz. It does this because as it moves faster, it needs more information about its surroundings in order to navigate. It uses the pulses like radar: to avoid obstacles and predators, to find food, and to locate other members of its species.

To enable an entire room to hear the electric fish, you will need an amplifier. A stereo system or boom box usually has an auxiliary input or phonograph input that you can use. Simply plug in a cable and place the two wires in the tank as you did for the earphone.

The experiment shown in the photograph on the following page used a Radio Shack battery-powered amplifier (part number 277-1008), which is small and easy to carry to a fish store.

A plug goes into the input jack, and two alligator clips connect the wires from the plug to two bare copper wires running into the water on the left and right sides of the tank. The fish has been temporarily placed in a tiny tank to make photographing it easier. Don't leave the fish there for more than an hour; this tank is much too small for long-term fish storage.

With the amplifier, an entire room can easily hear the clicks and buzzes as the fish reacts to the environment.

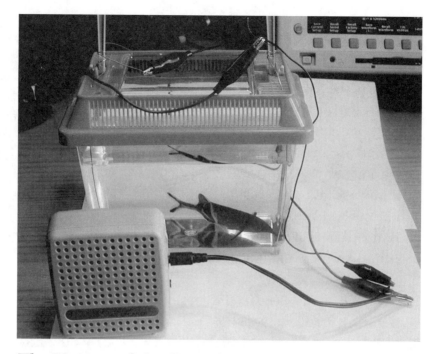

The Nature of the Signal

With an oscilloscope or a computer sound card, you can capture the signal and look at a picture of it. The graph shown here is a snapshot of a burst of speed by the fish as it darted around the tank.

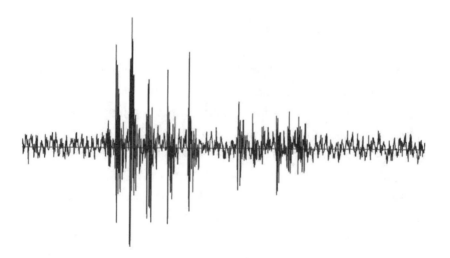

A close-up view of a single pulse is shown below.

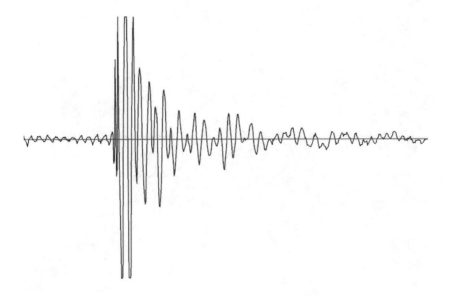

By simply putting a microphone up to the amplifier you can capture the sounds to a computer sound card and examine the graphs in detail. These graphs were captured with an expensive oscilloscope, and then graphed with a spreadsheet program on a computer, but you don't need expensive equipment to get similar results. A sound card is more than enough to get all the details at these frequencies.

The Electric Organ Discharge

The electric organ of the elephant nose fish is clearly visible. It is the narrow area connecting the skirt-like area of the fish to the tail fins.

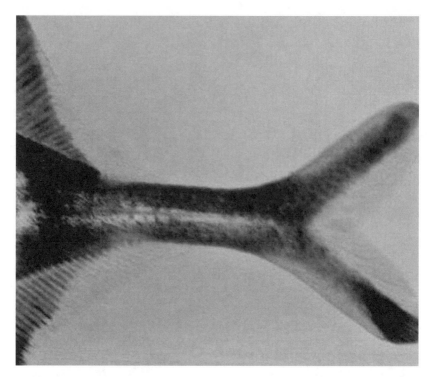

This is the organ that generates the signal. It is a modified muscle. To receive electric signals, either from other fish or from echoes of its own discharge, the elephant nose fish has three different types of receptors. These are the *mormyromasts*, the *knollenorgans*, and the *ampullary receptors*.

The mormyromasts are the organs that detect the echoes of the electric organ discharge. They are what enable the fish to navigate in murky water and find prey.

The knollenorgans detect the electric organ discharge of other elephant nose fish, and thus aid in communication and finding mates.

The ampullary receptors measure the low-frequency electric fields emitted by other aquatic animals. Similar ampullary receptors are used by catfish and sharks.

Fish are very sensitive to tiny amounts of pollutants in their environment. In some towns in Germany, the Elephant Nose fish is used to detect very small amounts of lead and trichloroethylene in the city's water supply. Because the electric discharges are so easy to detect and monitor with a computer, this method is cheaper than chemical tests and can be done continuously. The number of discharges per minute drops off measurably when the levels of either pollutant rises, even at levels far below those considered dangerous.

Other Electric Fish

Some of the better known electric fish are the South American electric eel *Electrophorus electricus* and the African electric catfish *Malopterurus electricus*. Another famous electric fish is the Mediterranean electric ray *Torpedo torpedo*. There is also an electric catfish from China, the *Parasilurus asota*. These are large fish with powerful electric discharge organs. They use the electric discharge to stun their prey and deter predators.

Other electric fish are more like the Elephant Nose fish, where the electric discharge is very small, and used for navigation and communication.

In the same family of fish (called mormyriform fish) as the elephant nose are several other species, such as *Pollimyrus isidori*, *Gymnarchus niloticus*, and *Brienomyrus brachyistius*.

Another family of weakly electric fish is the South American gymnotoid fish, such as *Hypopomus artedi*, *Sternopygus*, and *Eigenmannia*. While *Hypopomus* produces pulses like the Elephant Nose fish, the others produce pure continuous sine waves.

The South American gymnotoid fish *Eigenmannia virescens* (the Glass Knifefish), *Sternachella schotti* (the Brown Ghost Knifefish), and *Apteronotus albifrons* (the Black Ghost Knifefish) are easy to find in tropical fish stores. They are all of the continuous sine wave type, rather than the pulse type.

For more information on electric fish, enter any of the scientific names mentioned above into an Internet search engine. There is an amazing amount of material available on these animals on the Web.

Index

THE ART OF THE CATAPULT
Build Greek Ballistae, Roman Onagers, English Trebuchets, and More Ancient Artillery
By William Gurstelle

"A fascinating look at world history, military strategy, and physics, related with an engaging yet light-hearted touch."
—*School Library Journal*

"This book is a hoot . . . the modern version of *Fun for Boys* and *Harper's Electricity for Boys*." —*Natural History*

1556525265 · $16.95 (CAN $22.95)

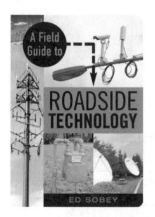

A FIELD GUIDE TO ROADSIDE TECHNOLOGY
By Ed Sobey

This fascinating handbook provides detailed descriptions and the history behind many of the devices that roadside travelers take for granted, from utility poles to satellite dishes. More than 150 roadside technologies are covered, and each detailed entry describes what the device does, how it works, and also includes a photograph for easy identification. Helpful sidebars describe related technical issues such as why stoplights are constructed with the red light on top.

1556526091 · $14.95 (CAN $20.95)

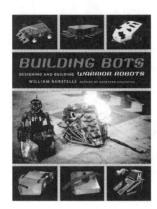